ABC OF CLINICAL GENETICS

Revised second edition

ABC OF CLINICAL GENETICS
REVISED SECOND EDITION

HELEN M KINGSTON
Consultant Clinical Geneticist
St Mary's Hospital, Manchester

BMJ
Publishing
Group

First published 1989
Second impression (revised) 1990
Second edition 1994
Second impression (revised) 1997

Third impression 1999
Fourth impression 2000

British Library Cataloguing in Publication Data

A catalogue record for this book is available from the British Library.

ISBN 0-7279-1101-5

Printed in Malaysia by Times Offset (M) Sdn Bhd
Typeset by Apek Typesetters Ltd, Nailsea, Bristol

Contents

ACKNOWLEDGMENTS

In writing this series I have been greatly influenced by my former teachers of clinical genetics, particularly Professors Peter Harper and Michael Laurence, to whom I am grateful. I have benefited from helpful suggestions on the text from Professors Rodney Harris, Dian Donnai and Andrew Read, Dr Gareth Evans, and Mr Roger Mountford of St Mary's Hospital, and thank many colleagues in Manchester who have generously provided illustrations for the series, Margaret Trickey for typing the original manuscripts, and Linda Nivern for typing the revised manuscript for the second edition.

CLINICAL GENETIC SERVICES

Genetic disease

Type of genetic disease

Single gene (mendelian)	Numerous though individually rare Clear pattern of inheritance High risk to relatives
Multifactorial	Common disorders No clear pattern of inheritance Low or moderate risk to relatives
Chromosomal	Mostly rare No clear pattern of inheritance Usually low risk to relatives
Somatic mutation	Accounts for mosaicism Cause of neoplasia

Genetic disorders place considerable health and economic burdens not only on affected people and their families but also on the community. As more environmental diseases are successfully controlled those that are wholly or partly genetically determined are becoming more important.

Despite a general fall in perinatal mortality rate the incidence of lethal malformations in newborn infants remains constant. Between 2–5% of all liveborn infants have genetic disorders or congenital malformations. These disorders have been estimated to account for one third of admissions to paediatric wards, and they contribute appreciably to paediatric mortality. Many common diseases in adult life also have a considerable genetic predisposition, including coronary heart disease, diabetes, and cancer.

Prevalence of genetic disease

Type of genetic disease	Estimated prevalence per 1000 population
Single gene:	
Autosomal dominant	2–10
Autosomal recessive	2
X linked recessive	1–2
Chromosomal abnormalities	6–7
Common disorders with appreciable genetic component	7–10
Congenital malformations	20
Total	38–51

Though diseases of wholly genetic origin are often individually rare, they are numerous and therefore important. Genetic disorders are incurable and often severe. A few are amenable to treatment, but most are not, so that emphasis is often placed on prevention of either recurrence within an affected family or complications in a person who is already affected.

Increasing awareness, both within the medical profession and in the general population, of the genetic contribution to disease has led to an increasing demand for clinical genetic services. Some aspects of genetics are well established and do not require referral to a specialised genetics clinic—for example, the provision of amniocentesis to exclude Down's syndrome in pregnancies at risk. Other aspects are less well understood by non-geneticists—for example, the role of molecular genetics in clinical practice, which is an area of rapidly advancing technology requiring the specialised facilities of a genetics centre.

Aims of genetic counselling

Genetic counselling covers more than estimating risks and extends beyond the person who presents to the whole family in changing situations over many years. The role of clinical geneticists is to establish an accurate diagnosis on which to base counselling and then to provide information about prognosis and follow up, the risk of developing or transmitting the disorder, and the ways in which this may be prevented or ameliorated. Throughout, the family requires support in adjusting to the implications of genetic disease and the consequent decisions that have to be made.

Clinical genetic services

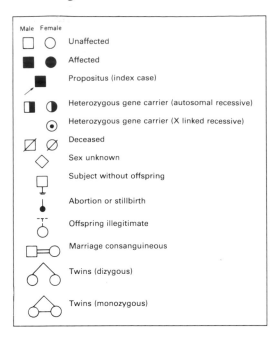

Male	Female	
□	○	Unaffected
■	●	Affected
■		Propositus (index case)
◨	◑	Heterozygous gene carrier (autosomal recessive)
	⊙	Heterozygous gene carrier (X linked recessive)
⊠	⊘	Deceased
◇		Sex unknown
⬚		Subject without offspring
●		Abortion or stillbirth
○		Offspring illegitimate
□—■		Marriage consanguineous
⋀		Twins (dizygous)
⬧		Twins (monozygous)

Diagnosis

An accurate diagnosis is the first essential requirement for genetic counselling. This may not always be straightforward as genetic disease is often variable in its presentation, and different members of a family with the same disorder may present to different specialties with diverse manifestations of the condition. Conversely, disorders which are clinically similar may follow different inheritance patterns in different families.

The person requesting genetic counselling may not be the one affected, and the diagnosis may need to be confirmed by examining the affected relative or reviewing their hospital records.

Without a defined diagnosis, appropriate genetic advice may be given if the pattern of affected subjects within a family points to a particular mode of inheritance.

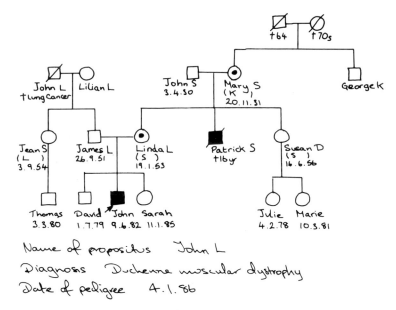

Name of propositus   John L
Diagnosis   Duchenne muscular dystrophy
Date of pedigree   4.1.86

Drawing a pedigree

Constructing a family tree is the best way to record genetic information. The main symbols used are shown in the upper box. It is important to record full names (including maiden names) and dates of birth on the pedigree. Specific questions should be asked about abortions, stillbirths, infant deaths, multiple marriages, and consanguinity as this information may not always be volunteered. It is also useful to record details of the medical care of relevant family members.

Common reasons for referral to a genetics clinic

Genetic disease diagnosed, counselling requested

Testing carrier state of family members for mendelian disorders

Investigation and diagnosis of possible genetic disease

Diagnosis of mental handicap or physical abnormality

Diagnosis of malformation in neonates or stillbirths

Genetic investigation of recurrent pregnancy loss

Genetic management of high risk pregnancies

Interpretation of abnormal prenatal tests

Estimation of risk

Estimation of genetic risk depends on the pattern of inheritance of a disorder and applies both to the risk of developing and of transmitting a particular disorder. In some disorders specific tests to identify carriers are available.

Mendelian disorders due to mutant genes generally carry high risks of recurrence whereas chromosomal disorders generally have low risks. For many common conditions there is no clearly defined pattern of inheritance, and empirical figures for the risk of recurrence are given, based on information derived from family studies.

Transmitting information

Interpretation of risk varies depending on the severity of the disorder, its prognosis, and the availability of treatment or palliation. All of these aspects need to be discussed with the family.

The risk of transmitting a disorder, the severity of the disorder, and the availability of prenatal diagnosis all influence the decisions of couples about pregnancy, as do their moral and religious convictions. Contraception or sterilisation may be considered, and alternative options may include insemination by a donor, ovum donation, or adoption.

It is important that the counselling process is not directive and that couples can reach their own decisions armed with the necessary information.

Psychological aspects

The diagnosis of genetic disease causes considerable emotional stress, and to be effective the counselling process must provide psychological support in addition to information. Recognition of the impact of genetic disease and the various stages of the process of coping allows counselling to be pursued at an appropriate pace for each couple. Some knowledge of the couple's educational, social, and religious backgrounds is important as these influence their reactions and decision making.

Counselling must be unhurried and undertaken in a quiet environment. The counsellor needs to spend sufficient time with the couple to establish mutual rapport, so that personal feelings can be freely discussed and questions asked and dealt with sensitively; several counselling sessions, either in the clinic or at the patient's home may be necessary to achieve this.

Main users of clinical genetic services

Paediatricians

Obstetricians

Other hospital specialists

General practitioners

Community child health services

Others (self referrals, adoption services)

Organisation of clinical services

Medical staff	Consultant clinical geneticists
	Specialist registrars
	Clinical assistants or clinical medical officers
Field workers	Genetic associates (scientific officers or counsellors)
	Specialist health visitors
	Social workers
	Phlebotomist
Clerical staff	

Departments of clinical genetics tend to be based regionally in main teaching centres and often have academic as well as NHS staff. Clinical services are often provided in district based clinics by staff from the regional centre to ensure equal access to genetic services for patients and families unable to travel to the main centre. Specialty clinics are often provided, focusing on specific disorders such as birth defect syndromes, cystic fibrosis, familial cancer, Huntington's disease and muscular dystrophy.

Diagnosis and genetic counselling is provided by medically qualified clinical geneticists. There is an increasing involvement in counselling by trained non-medical genetic counsellors who often have a nursing or scientific background and qualification in counselling.

Associated laboratory services

Specialist laboratory services form an integral part in providing clinical genetic services. The laboratories are usually based in regional or supraregional centres and provide services in biochemical genetics, cytogenetics, and molecular genetics.

Genetic registers

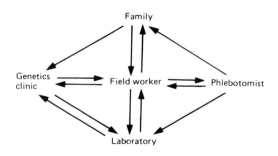

Clinical use of genetic registers is aimed primarily at ascertaining as completely as possible all people at risk of developing or transmitting a particular disorder so that appropriate counselling can be offered. A register approach permits long term follow up of family members, which is important for children at risk, who will not need investigation or counselling for many years. Registers are particularly useful for disorders that are amenable to DNA analysis in which advances are of clinical importance and families need regular counselling with new information. Disorders suited to a register approach include dominant disorders with late onset, such as Huntington's disease and myotonic dystrophy, and X linked disorders, such as Duchenne and Becker's muscular dystrophy. Registers can also provide data on the incidence and natural course of diseases and the effect of counselling and preventive programmes.

Genetic registers are held on computer and are subject to the Data Protection Act. No one is included in a register without having given informed consent.

MENDELIAN INHERITANCE

Gregor Mendel 1822–84

Disorders caused by a defect in a single gene follow the patterns of inheritance described by Mendel. Individual disorders of this type are often rare but are important because they are numerous (over 4000 single gene traits have been listed[1]). Risks within an affected family are usually high and are calculated by knowing the mode of inheritance and details of the family pedigree.

Autosomal dominant disorders

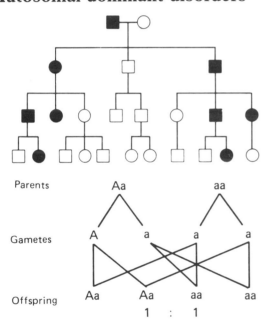

Parents	Aa		aa
Gametes	A a	a a	
Offspring	Aa Aa	aa aa	
	1	:	1

Autosomal dominant disorders affect both males and females and can often be traced through many generations of a family. Affected people are heterozygous for the abnormal allele and transmit the gene for the disease to half their offspring, whether male or female. The disorder is not transmitted by family members who are unaffected themselves. Estimation of risk is therefore apparently simple, but in practice several factors may cause difficulties in counselling families.

Firstly, the age of onset of a disorder may be variable and people with a defective gene, who are destined to become affected, may remain without signs or symptoms well into adult life. Young people at risk may not know whether they have inherited the disorder and will transmit it to their children at a time when they are planning their own families. Detection of people carrying the mutant gene before symptoms become apparent may therefore be important in conditions such as Huntington's disease and myotonic dystrophy.

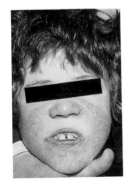

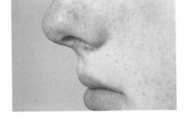

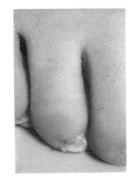

The severity of many dominant conditions also varies considerably among affected members within a family. The likely severity in any affected offspring is difficult to predict, and a mildly affected parent may have a severely affected child, as illustrated by tuberous sclerosis, in which a parent with only skin manifestations of the disorder may have an affected child with infantile spasms and severe mental retardation.

New mutation may account for the presence of a dominant disorder in a subject who does not have a family history of the disease. When a disorder arises by new mutation the risk of recurrence in future pregnancies for the parents of the affected child is negligible. Care

Tuberous sclerosis. Top left: severely affected boy with fits and mental retardation; top right: adenoma sebaceum; bottom left: ash leaf depigmentation; bottom right: periungal fibroma.

Examples of autosomal dominant disorders

Achondroplasia
Acute intermittent porphyria
Adult polycystic kidney disease
Alzheimer's disease (some cases)
Epidermolysis bullosa (some forms)
Facioscapulohumeral dystrophy
Familial hypercholesterolaemia

Huntington's disease
Myotonic dystrophy
Noonan's syndrome
Neurofibromatosis
Osteogenesis imperfecta (some forms)
Familial adenomatous polyposis
Tuberous sclerosis

must be taken to exclude a mild form of the condition in one or other parent before giving this reassurance. New mutation accounts for most cases of achondroplasia, a condition that can be easily excluded in the parents. On the other hand, neurofibromatosis may arise by new mutation or be present in mild form in one parent. In dominant conditions an apparently normal parent may occasionally carry a germline mutation; this is associated with a considerable risk of recurrence. A dominant disorder in a person with a negative family history may alternatively indicate non-paternity.

A few dominant disorders show lack of penetrance—that is, a person who inherits the gene does not develop the disorder. In this case people who are not affected cannot be completely reassured that they will not transmit the disorder to their children. The risk is, however, fairly low, not exceeding 10%, because when penetrance is high an unaffected person is unlikely to be a gene carrier, and when it is low the chance of a gene carrier developing the disorder is correspondingly small.

Non-genetic factors may also influence the expression of dominant genes—for example, diet in hypercholesterolaemia and drugs in porphyria.

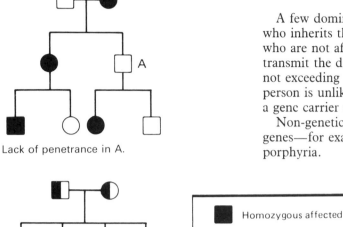

Lack of penetrance in A.

Homozygosity for dominant genes is uncommon, unless two people with the same disorder marry. This may happen preferentially with certain conditions, such as achondroplasia. Homozygous achondroplasia is a lethal condition and the risks for offspring are therefore: 25% homozygous affected (lethal); 50% heterozygous affected; 25% homozygous normal. Thus two out of three living children will be affected.

■ Homozygous affected
◨ Heterozygous affected

Homozygosity for a dominant disorder.

Autosomal recessive disorders

Autosomal recessive disorders occur in a person whose healthy parents both carry the same recessive gene. The risk of recurrence for future offspring of such parents is 25%. Unlike autosomal dominant disorders there is generally no family history. Although the defective gene may be passed from generation to generation, the disorder generally only appears within a single sibship—that is, within one group of brothers and sisters.

In northern Europeans the commonest autosomal recessive disorder is cystic fibrosis, and about one in 20 people in the population is a carrier.

Consanguinity increases the risk of a recessive disorder because both parents are more likely to carry the same defective gene, which has been inherited from a common ancestor. The rarer the condition the more likely it is that the parents were related before marriage. Overall, the increased risk to parents who are first cousins of having a child with severe abnormalities is fairly low (3% above the risk in the general population), and this includes the risk of autosomal recessive disorders.

The offspring of an affected person will be healthy heterozygotes and can be affected only if the other parent is also a gene carrier. This is unlikely except in consanguineous marriages or in ethnic groups in which particular genes are common.

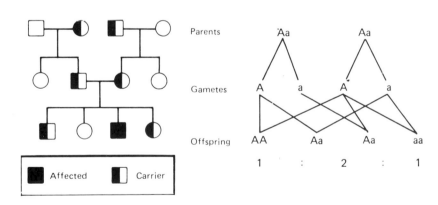

	Parents	Aa	Aa
	Gametes	A a	A a
	Offspring	AA Aa Aa aa
		1 : 2 : 1

■ Affected ◨ Carrier

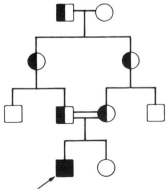

Consanguinity and autosomal recessive inheritance.

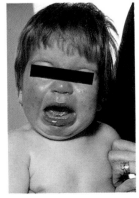

Hurler's syndrome: coarsening of facial features.

Examples of autosomal recessive disorders

Congenital adrenal hyperplasia	Homocystinuria
Cystic fibrosis	Hurler's syndrome (mucopolysaccharidosis I)
Deafness (some forms)	Laurence-Moon-Biedl syndrome
Diastrophic dwarfism	Occulocutaneous albinism
Epidermolysis bullosa (some forms)	Phenylketonuria
Friedreich's ataxia	Sickle cell disease
Galactosaemia	Tay-Sachs disease
Haemochromatosis	Thalassaemia

Autosomal recessive disorders are commonly severe, and many of the recognised inborn errors of metabolism follow this type of inheritance. Many complex malformation syndromes are also due to autosomal recessive genes, and their recognition is important in the first affected child in a family because of the 25% risk of recurrence. Prenatal diagnosis for recessive disorders may be possible by performing biochemical assays, DNA analysis, or looking for structural abnormalities in the fetus.

X linked recessive disorders

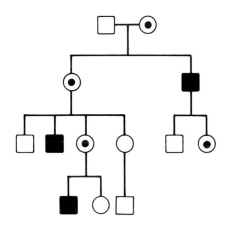

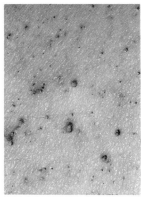

Angiokeratoma in Fabry's disease.

In X linked recessive conditions only males are affected, but the disorder can be transmitted through healthy female carriers.

A female carrier of an X linked recessive disorder will transmit the condition to half her sons, and half her daughters will be carriers. An unaffected male does not transmit the disorder. An affected male will transmit the mutant gene to all his daughters (who must inherit his X chromosome), but to none of his sons (who must inherit his Y chromosome). This absence of male to male transmission is a hallmark of X linked inheritance. Many X linked recessive disorders are severe or lethal during early life, however, so that the affected males do not reproduce.

Occasionally a heterozygous female will show some features of the condition. This is usually due to non-random X inactivation leading to the chromosome that carries the mutant allele remaining active in most cells. Alternatively, X chromosome abnormalities such as Turner's syndrome may give rise to X linked disorders in females. Homozygous affected state may occur in females whose father is affected and whose mother is a carrier. However, this is only likely to occur in common X linked disorders such as red-green colour blindness, or glucose-6-phosphate dehydrogenase deficiency in the Middle East.

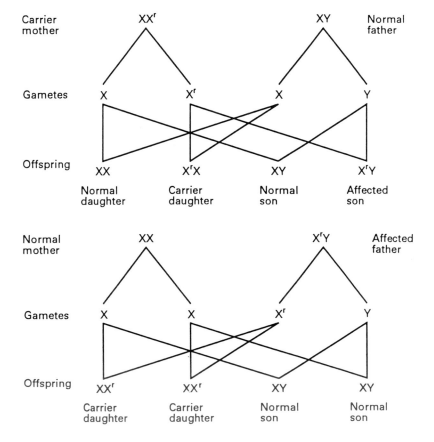

An X linked recessive condition should be considered when the family history indicates maternally related affected males in different generations of the family. Family history is not always positive, however, as new mutations are fairly common. Recognising X linked recessive inheritance is important because many female relatives may be at risk of being carriers and of having affected sons, irrespective of whom they marry.

Examples of X linked disorders

Recessive

Anhidrotic ectodermal dysplasia

Becker's muscular dystrophy

Colour blindness

Duchenne muscular dystrophy

Fabry's disease

Glucose-6-phosphate dehydrogenase deficiency

Haemophilia A, B

Hunter's syndrome (mucopolysaccharidosis II)

Lesch-Nyhan syndrome

Menkes's syndrome

Mental retardation with or without fragile site

Ocular albinism

Dominant

Incontinentia pigmenti

Orofaciodigital syndrome

Rickets resistant to vitamin D

Identifying female gene carriers requires interpretation of the family pedigree and the results of specific tests to identify carriers. This is not always straightforward as some women whose results are normal will carry germline mutations not amenable to detection and be at risk of having affected sons.

Male with fragile X syndrome (X linked mental retardation)

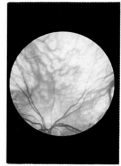

X linked ocular albinism

X linked dominant disorders

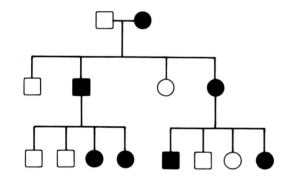

An X linked dominant gene will give rise to a disorder in both hemizygous males and heterozygous females. The gene is transmitted in families in the same way as X linked recessive genes, giving rise to an excess of affected females. In some disorders the condition is lethal in hemizygous males. In this case there will be fewer males than expected in the family, all of whom will be healthy, and an excess of females, half of whom will be affected.

Y linked disorders

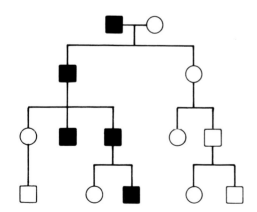

In Y linked disorders only males are affected, with transmission being directly from father to son with the Y chromosome. This pattern of inheritance has been suggested for such conditions as porcupine skin, hairy ears, and webbed toes. In most conditions in which Y linked inheritance has been postulated the actual mode of inheritance is probably autosomal dominant, with other factors causing sex limitation.

1 McKusick VA. *Mendelian inheritance in man. Catalogs of autosomal dominant, autosomal recessive and X-linked phenotypes.* 8th ed. Baltimore: Johns Hopkins, 1988.

The illustrations of tuberous sclerosis were reproduced by kind permission of Professor P S Harper, Institute of Medical Genetics for Wales, Cardiff.

UNUSUAL INHERITANCE MECHANISMS

Unstable mutations

> **Examples of single gene disorders due to trinucleotide repeat expansion**
>
> Myotonic dystrophy
>
> Fragile X syndrome
>
> Friedreich's ataxia
>
> Huntington's disease
>
> Spinobulbar neuronopathy (Kennedy syndrome)
>
> Spinocerebellar ataxia 1

The recent discovery of an unstable mutation mechanism involving a trinucleotide repeat has identified a new cause of some human genetic diseases that are inherited in mendelian fashion. The three common disorders due to trinucleotide repeat expansions—myotonic dystrophy, fragile X syndrome, and Huntington's disease—all share unusual characteristics of inheritance that can be explained by the instability of the mutation. These include the phenomenon of anticipation, in which the disorder becomes more severe in successive generations of a family, and the striking parental sex bias in the inheritance of the most severe forms of the disorder (maternal in myotonic dystrophy and fragile X syndrome and paternal in Huntington's disease). Instability of the mutation usually generates larger expansions in offspring, but reduction in size of the expansion is also documented. Trinucleotide repeat expansions are discussed further in the chapters on DNA analysis in genetic disorders and on molecular genetics of selected mendelian disorders.

Imprinting

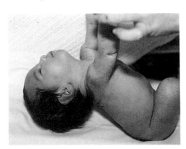

Severe hypotonia in infant with Prader-Willi syndrome.

It has been observed that some inherited traits do not conform to the pattern expected of classical mendelian inheritance in which genes inherited from either parent have an equal effect. The term imprinting is used to describe the phenomenon by which certain genes function differently, depending on whether they are maternally or paternally derived. The mechanism of DNA modification involved in imprinting remains to be explained, but it confers a functional change in particular alleles at the time of gametogenesis determined by the sex of the parent. The imprint lasts for one generation and is then removed, so an appropriate imprint can be re-established in the germ cells of the next generation.

Ataxic gait in child with Angelman's syndrome.

The effects of imprinting can be observed at several levels: that of the whole genome, that of particular chromosomes or chromosomal segments, and that of individual genes. For example, the effect of triploidy in human conceptions depends on the origin of the additional haploid chromosome set. When paternally derived, the placenta is large and cystic with molar changes and the fetus has a large head and small body. When the extra chromosome set is maternal, the placenta is small and underdeveloped without cystic changes and the fetus is noticeably underdeveloped.

One of the best examples of imprinting in human disease is shown by deletions in the q11–13 region of chromosome 15, which may cause either Prader-Willi syndrome or Angelman's syndrome. The features of Prader-Willi syndrome are severe neonatal hypotonia and failure to thrive with later onset of obesity, behaviour problems, mental retardation, characteristic facial appearance, small hands and feet, and hypogonadism. Angelman's syndrome is quite distinct and is associated with severe mental retardation, microcephaly, ataxia, epilipsy, and absent speech.

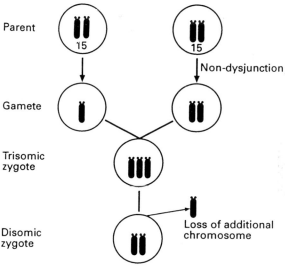

Parent

Gamete

Trisomic
zygote

Disomic
zygote

Non-dysjunction

Loss of additional
chromosome

Loss of one chromosome 15, thus restoring
normal chromosome number to an initially
trisomic zygote. The two chromosome 15s
remaining could represent one chromosome from
each parent or uniparental disomy, depending on
which chromosome is lost during the process of
trisomy rescue.

The genes for Prader-Willi and Angelman's syndromes are both situated within the 15q11–13 region. Similar de novo cytogenetic or molecular deletions can be detected in both conditions. In Prader-Willi syndrome the deletion always occurs on the paternally derived chromosome 15, whereas the deletion in Angelman's syndrome is always on the maternally derived chromosome. In patients with Prader-Willi syndrome who do not have a chromosome deletion, both chromosome 15s are maternally derived. This phenomenon is called uniparental disomy and, when involving imprinted regions of the genome it has the same effect as a chromosomal deletion arising from the opposite parental chromosome. In Prader-Willi syndrome both isodisomy (inheritance of identical chromosome 15s from one parent) and heterodisomy (inheritance of different 15s from the same parent) have been observed. The origin of this uniparental disomy is probably the loss of one chromosome 15 from a conception that was initially trisomic. Uniparental disomy is rare in Angelman's syndrome, but when documented has involved disomy of the paternal chromosome 15.

Imprinting has been implicated in other human diseases, notably in some forms of cancer, such as familial glomus tumours, Wilm's tumours, and Beckwith-Wiedemann syndrome, which predisposes to Wilm's tumour.

Mosaicism

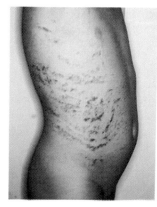

Patchy distribution of skin lesions in female with incontinentia pigmenti, an X linked dominant disorder, lethal in males but not in females, because of functional X chromosomal mosaicism

Mosaicism refers to the presence of two or more cell lines, which differ in chromosomal constitution or genotype but have been derived from a single zygote. Mosaicism is a postzygotic event that may arise in early embryonic development, or in later fetal or postnatal life. The time at which the mosaicism develops will determine the relative proportions of the two cell lines, and hence the severity of abnormality of the phenotype caused by the abnormal cell line. Functional mosaicism occurs in females as one X chromosome remains active in each cell. The process of X inactivation occurs in early embryogenesis and is random. Thus, alleles that differ between the two chromosomes will be expressed in mosaic fashion. Carriers of X linked recessive mutations normally remain asymptomatic as only a proportion of cells have the mutant allele on the active chromosome. Occasional females will, by chance, have the normal X chromosome inactivated in the majority of cells and will then manifest symptoms of the disorder caused by the mutant gene.

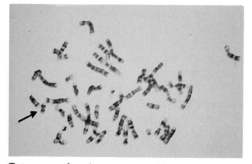

Tetrasomy for chromosome 12p occurs only in mosaic form in liveborn infants

Chromosomal mosaicism is not infrequent, and arises by postzygotic errors in mitosis. Mosaicism is observed in Turner's syndrome and Down's syndrome, and the phenotype is less severe than in cases with complete aneuploidy. Mosaicism has been documented for many other numerical or structural chromosomal abnormalities that would be lethal in non-mosaic form. The clinical importance of chromosomal mosaicism detected prenatally may be difficult to assess. The abnormal karyotype detected by amniocentesis or chorionic villus sampling may be confined to placental cells, but even when present in the fetus the severity with which the fetus will be affected is difficult to predict.

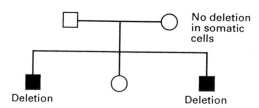

No deletion in somatic cells

Deletion

Deletion

Pedigree showing recurrence of Duchenne muscular dystrophy because of dystrophin gene deletion in the sons of a woman who does not carry the deletion in her leucocyte DNA. Recurrence is caused by gonadal mosaicism, in which the mutation is confined to some of the germline cells in the mother.

Single gene mutations occurring in somatic cells also result in mosaicism. In mendelian disorders this may present as a patchy phenotype, as in segmental neurofibromatosis type I. Somatic mutation is also a mechanism for neoplastic change.

Germline mosaicism is one explanation for the transmission of a genetic disorder to more than one offspring by apparently normal parents. In Duchenne muscular dystrophy, it has been calculated that up to 20% of the mothers of isolated cases, whose carrier tests give normal results, may have gonadal mosaicism for the muscular dystrophy mutation. The possibility of germline mosaicism makes it difficult to exclude a risk of recurrence in X linked recessive and autosomal dominant disorders, for parents who are apparently normal.

Mitochondrial disorders

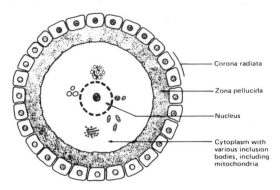

Diagrammatic representation of human egg.

Not all DNA is contained within the cell nucleus. Mitochondria have their own DNA consisting of a double stranded circular molecule. This mitochrondrial DNA consists of 16 567 base pairs that constitute 37 genes. There is some difference in the genetic code between the nuclear and mitochondrial genomes, and mitochondrial DNA is almost exclusively coding, with the genes containing no intervening sequences (introns). A diploid cell contains two copies of the nuclear genome, but there may be more than 1000 copies of the mitochondrial genome, as each mitochondrium contains up to 10 copies of its circular DNA and each cell contains hundreds of mitochondria. Mutations within mitochondrial DNA appear to be 5 or 10 times more common than mutations in nuclear DNA, and the accumulation of mitochondrial mutations with time has been suggested as having a role in ageing.

Examples of diseases caused by mitochondrial DNA mutations

Disorder	Symptoms	Mutation	Inheritance
Leber's hereditary optic neuropathy (Leber's optic atrophy)	Acute visual loss and possibly other neurological symptoms	Point mutation at position 11778 in ND4 gene of complex I	Maternal
MERRF	Myoclonic epilepsy with other neuro-logical symptoms and ragged red fibres in skeletal muscle	Point mutation in tRNA lys gene	Sporadic, occasionally maternal
Kaerns-Sayre syndrome	Progressive external ophthalmoplegia, pigmentary retinopathy, heart block, ataxia, muscle weakness, deafness	Large tandem duplication / Large deletion	Sporadic / Usually sporadic
MELAS	Encephalomyopathy, lactic acidosis, stroke-like episodes	Point mutation in tRNA leu gene	Sporadic, occasionally maternal

The mitochondrial genome encodes 22 types of transfer and ribosomal RNA molecules that are involved in mitochondrial protein synthesis, as well as 13 of the polypeptides involved in the respiratory chain system. The remaining respiratory chain polypeptides are encoded by nuclear genes. Diseases affecting mitochondrial function may therefore be controlled by nuclear gene mutation and follow mendelian inheritance, or may result from mutations within the mitochondrial DNA. As the main function of mitochondria is the synthesis of ATP by oxidative phosphorylation, mitochondrial disorders are most likely to affect tissues such as the brain, skeletal muscle, cardiac muscle, and eye which contain abundant mitochondria and rely on aerobic oxidation and ATP production.

Mutations in mitochondrial DNA have been identified in a number of diseases, notably Leber's hereditary optic neuropathy, MERRF (myoclonic epilepsy with ragged red fibres), MELAS (mitochondrial myopathy with encephalopathy, lactic acidosis, and stroke-like episodes), and progressive external ophthalmoplegia including Kaerns-Sayre syndrome. Disorders due to mitochondrial mutations often appear to be sporadic. When they are inherited, however, they demonstrate maternal transmission. This is because only the egg contributes cytoplasm and mitochondria to the zygote. All offspring of a carrier mother will carry the mutation, all offspring of a carrier father will be normal. The pedigree pattern in mitochondrial inheritance may be difficult to recognise, however, because some carrier individuals remain asymptomatic. In Leber's hereditary optic neuropathy for example, half the sons of a carrier mother would be affected, but only 1 in 5 of the daughters will be symptomatic. Nevertheless, all daughters would transmit the mutation to their offspring. The descendants of case fathers are never affected.

KEY ■●Clinically affected ▨◉Carriers of mitochondrial mutation

Pedigree of Leber's hereditary optic neuropathy caused by a mutation within the mitochondrial DNA. Carrier women transmit the mutation to all their offspring, some of whom will develop the disorder. Affected or carrier men do not transmit the mutation to any of their offspring.

Genetic counselling dilemmas in mitochondrial diseases

- Some disorders of mitochondrial function are due to nuclear gene mutations
- Many disorders caused by mitochondrial mutations are sporadic
- It is not known whether the degree of heteroplasmy in the mother determines risk to offspring
- Severity is very variable and difficult to predict
- It is difficult to advise asymptomatic relatives who carry the mitochondrial mutation

Because multiple copies of mitochondrial DNA are present in the cell, mitochondrial mutations are often heteroplasmic—that is, a single cell will contain a mixture of mutant and wild-type mitochondrial DNA. With successive cell divisions some cells will remain heteroplasmic but others may drift towards homoplasmy for the mutant or wild-type DNA. Large deletions, which make the remaining mitochondrial DNA appreciably shorter, may have a selective advantage in terms of replication efficiency, so that the mutant genome accumulates preferentially. The severity of disease caused by mitochondrial mutations probably depends on the relative proportions of wild-type and mutant DNA present, but is very difficult to predict in a given subject.

The illustrations of incontinentia pigmenti and tetrasomy 12p were reproduced by kind permission of Professor D Donnai of St Mary's Hospital, Manchester.

ESTIMATION OF RISK

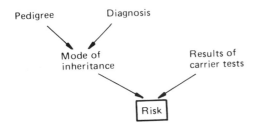

This chapter gives some examples of the risk of developing or transmitting a mendelian disorder. The mathematical risk calculated from data on a pedigree may often be modified by additional information from specific tests to detect carriers. Risk can be expressed as either a percentage or a fraction.

Autosomal dominant disorders

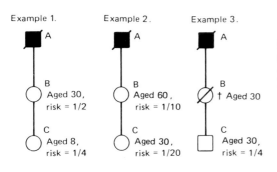

The 50% risk of developing a condition for the offspring of an affected person may be modified by age in disorders whose onset is in adult life, such as Huntington's disease. In examples 1 and 2 the risk to person B of developing Huntington's disease is still 50% at age 30 years, but by the age of 60 the residual risk to a healthy person has fallen to about 10%. The risk to person C therefore falls from 25% in example 1 to 5% in example 2. In example 3 the risk for C cannot be reduced below 25% because parent B, although clinically unaffected, died aged 30 while still at 50% risk.

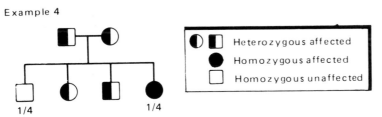

When both parents have the same autosomal dominant disorder, the risk to the offspring will be high (example 4). Only one in four children will be unaffected, and one in four will be homozygous for the mutant gene, which may cause severe disease, as in familial hypercholesterolaemia or achondroplasia.

Reduced penetrance also modifies simple autosomal dominant risk. Example 5 shows the risks for a disorder with 80% penetrance in which only 80% of gene carriers develop the disorder. Although clinically unaffected, person B may still carry the mutant gene, and there is therefore a fairly small risk of her child developing the disorder.

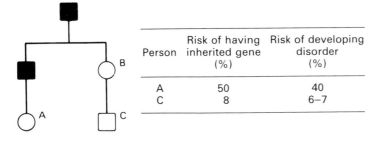

Person	Risk of having inherited gene (%)	Risk of developing disorder (%)
A	50	40
C	8	6–7

Autosomal recessive disorders

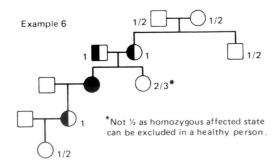

*Not ½ as homozygous affected state can be excluded in a healthy person.

Recurrence of autosomal recessive disorders generally only occurs within a particular sibship. Many members of the family, however, may be gene carriers, the risks of which are shown in example 6.

Example 7

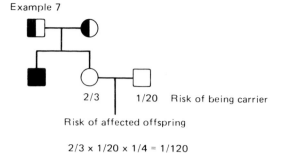

2/3   1/20   Risk of being carrier

Risk of affected offspring

2/3 x 1/20 x 1/4 = 1/120

The chance of a healthy sibling having affected children is low. The actual risk depends on the frequency of the gene in the general population. The risk for cystic fibrosis is shown in example 7. (In the general population about one in 20 people are gene carriers.)

Example 8

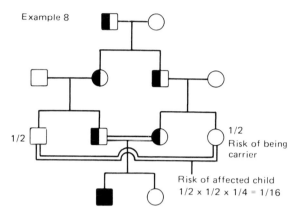

1/2

1/2
Risk of being carrier

Risk of affected child
1/2 x 1/2 x 1/4 = 1/16

When there is a tradition of consanguinity more than one marriage may be arranged between two families. If a consanguineous couple have a child affected by an autosomal recessive condition other family marriages may also be at risk of having affected offspring, as in example 8. The risk may not be high enough to prevent further planned marriages taking place, but if carrier state can be determined by specific tests this will help the families to make decisions and reassure the relatives who are not carriers.

Example 9   Example 10

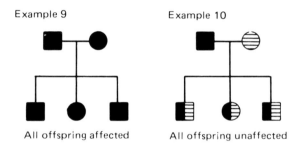

All offspring affected   All offspring unaffected

When both parents are affected by a recessive condition such as deafness the risks to the offspring will depend on whether the parents are homozygous for allelic or non-allelic genes as some autosomal recessive disorders can be caused by different genes at separate loci. In example 9 both parents have the same form of recessive deafness and all their children will be affected. In example 10 the parents have different forms of recessive deafness due to genes at different loci. Their offspring will be heterozygous at both loci and therefore unaffected.

X linked recessive disorders

Example 11

Example 12

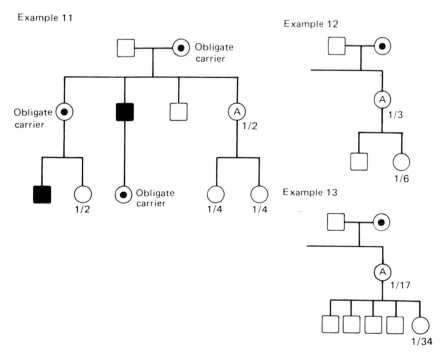

Example 13

Calculation of risks in X linked recessive disorders is often complex, and the examples opposite illustrate some basic concepts without details of method. Referral to a specialist genetic centre is usually indicated for calculating carrier state, which depends on pedigree structure and the results of specific tests. In families with X linked recessive disorders many female relatives are at risk of being carriers, as in example 11.

If the female relative A at risk in example 11 has any healthy sons this will reduce her risk to the values shown in examples 12 and 13.

Estimation of risk

Example 14

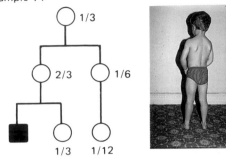

In lethal X linked recessive disorders new mutations account for a third of all cases. A mother of an affected boy is therefore not always a carrier. Carrier risks in families in which there is an isolated case of such a disorder (for example, Duchenne muscular dystrophy) are shown in example 14. These risks would again be modified by the presence of any unaffected males in the pedigree as well as by the results of specific tests to detect carriers.

Winging of scapulae, exaggerated lumbar lordosis, and prominent calf muscles in boy with Duchenne muscular dystrophy.

Isolated cases

Example 15

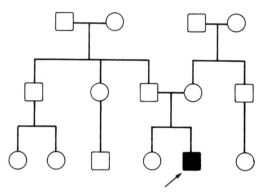

Pedigrees showing only one affected person are the type most commonly encountered in clinical practice (example 15). Various causes must be considered, and counselling in this situation depends entirely on reaching an accurate diagnosis in the affected person.

Possible causes

- Non-genetic
- Autosomal dominant: new mutation, non-paternity, or (rarely) parental germline mutation
- Autosomal recessive
- X linked recessive. May represent new mutation or inheritance from carrier mother
- Polygenic (multifactorial). Risk of recurrence generally low
- Chromosomal. Recurrence depends on type of abnormality, but generally low

---

Example 16

Risk of recurrence
$7/10 \times 2/3 \times 1/4 \hat{=} 1/9$

In example 16 calculation of the risk of recurrence after an isolated case of congenital deafness is based on the finding that when environmental causes are excluded 70% of cases are genetic, of which two thirds are autosomal recessive.

DETECTION OF CARRIERS

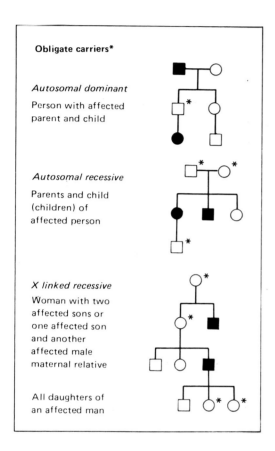

Obligate carriers*

Autosomal dominant

Person with affected
parent and child

Autosomal recessive

Parents and child
(children) of
affected person

X linked recessive

Woman with two
affected sons or
one affected son
and another
affected male
maternal relative

All daughters of
an affected man

Identifying carriers of genetic disorders in families or populations at risk plays an important part in preventing genetic disease. A carrier is a healthy person who possesses the mutant gene for an inherited disorder in the heterozygous state. The term carrier is therefore restricted to people at risk of transmitting mendelian disorders and does not apply to parents whose children have chromosomal abnormalities or congenital malformations such as neural tube defect. An exception is that people who have balanced chromosomal translocations are referred to as carriers as the inheritance of balanced or unbalanced translocations follows mendelian principles.

In families in which there is a genetic disorder some members must be carriers because of the way in which the condition is inherited. These obligate carriers can be identified by drawing a family pedigree and do not require testing as their genetic state is not in doubt. Identification of obligate carriers is important not only for their counselling but also for defining a group in whom tests for carrier state can be evaluated. Knowledge is needed of the proportion of obligate carriers showing abnormalities on clinical examination or with specific investigations and of the age at which such abnormalities appear to assess the likelihood of carrier state in the other relatives.

In some disorders—for example, sickle cell disease—all carriers can be identified with certainty; in others—for example, tuberous sclerosis—only a proportion can be identified. In autosomal dominant and X linked recessive disorders parental carrier state may be particularly difficult to assess because of the additional possibilities of new mutation in the child or germline mosaicism in the parent.

Autosomal dominant disorders

Some autosomal dominant disorders amenable to carrier detection

Adult polycystic kidney disease

Familial hypercholesterolaemia

Huntington's disease

Malignant hyperthermia

Myotonic dystrophy

Neurofibromatosis

Tuberous sclerosis

von Hippel-Lindau disease

In autosomal dominant conditions most heterozygous subjects are clinically affected and testing for carrier state applies only to disorders that are either variable in their manifestations or have a late onset. Gene carriers in conditions such as tuberous sclerosis may be mildly affected but run the risk of having severely affected children whereas carriers in other disorders, such as Huntington's disease, are destined to develop severe disease themselves.

Identifying symptomless gene carriers allows a couple to make informed decisions about having children, may indicate a need to avoid environmental triggers (as in porphyria), or may permit early treatment and prevention of complications (for example, in von Hippel-Lindau disease). Although testing for carrier state can have important benefits in conditions in which the prognosis is improved by early detection, presymptomatic diagnosis of severe disorders, such as Huntington's disease, that are not amenable to treatment presents problems. Exclusion of carrier state is, however, equally important, removing anxiety about transmitting the condition to offspring and the need for long term follow up.

Detection of carriers

Autosomal recessive disorders

In autosomal recessive conditions carriers remain healthy. Occasionally, heterozygous subjects may show minor abnormalities, such as altered red cell morphology in sickle cell disease and mild anaemia in thalassaemia. Most inborn errors of metabolism follow autosomal recessive inheritance, and heterozygous subjects may show reduced activities of specific enzymes, which provides the basis for detecting carriers.

The parents of an affected child are obligate carriers, but testing may be appropriate for the healthy siblings of an affected person and their partner if the condition is fairly common. Testing may also be important for consanguineous couples with a positive family history of genetic disease. The main opportunity for preventing autosomal recessive disorders, however, depends on population screening programmes, which will identify couples at risk before the birth of an affected child within the family. Screening subgroups of the population at high risk has proved effective in Tay-Sachs disease and β thalassaemia and is applicable to cystic fibrosis now that the common mutations can be tested for.

X linked recessive disorders

Some X linked recessive disorders amenable to carrier detection

Albinism (ocular)
Angiokeratoma (Fabry's disease)
Chronic granulomatous disease
Ectodermal dysplasia (anhidrotic)
Fragile X syndrome
Haemophilia A
Haemophilia B
Ichthyosis (steroid sulphatase deficiency)
Lesch-Nyhan syndrome
Menkes's syndrome
Mucopolysaccharidosis II (Hunter's syndrome)
Muscular dystrophy (Duchenne and Becker's)
Ornithine transcarbamylase deficiency
Retinitis pigmentosa

Carrier detection in X linked recessive conditions is particularly important as these disorders are often severe and in an affected family many female relatives may be at risk of having affected sons, irrespective of whom they marry. Genetic counselling cannot be undertaken without accurate assessment of carrier state, and calculating the risk is often complex.

Obligate carriers do not always show abnormalities on biochemical testing because of lyonisation, a process by which one or other X chromosome in female embryos is randomly inactivated early in embryogenesis. The proportion of cells with the normal or mutant X chromosome remaining active varies and will influence detection of carrier state. Carriers with a high proportion of normal X chromosomes remaining active will show no abnormalities on biochemical testing. Conversely, carriers with a high proportion of mutant X chromosomes remaining active are more likely to show biochemical abnormalities and may occasionally develop signs and symptoms of the disorder. Females with symptoms are called manifesting carriers.

Overlapping ranges of serum creatine kinase activity in controls and obligate carriers of Becker's muscular dystrophy. (Ranges vary among laboratories.)

Biochemical tests designed to determine carrier state must be evaluated initially in obligate carriers identified from affected families. Only tests which give significantly different results in obligate carriers compared with controls will be useful in determining the genetic state of female subjects at risk. Because the ranges of values in obligate carriers and controls overlap considerably (for example serum creatine kinase activity in X linked muscular dystrophy) the results for possible carriers are expressed in relative terms as a likelihood ratio. With this type of test confirmation of carrier state is always easier than exclusion. In muscular dystrophy a high serum creatine kinase activity confirms the carrier state; a normal result reduces but does not eliminate the chance that a female is a carrier.

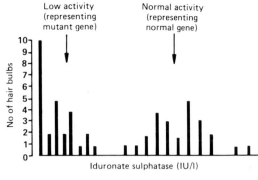

The problem of lyonisation can be largely overcome if biochemical tests can be performed on clonally derived cells; hair bulbs have been successfully used to detect carriers of Hunter's syndrome (mucopolysaccharidosis II). Carriers can be identified because they have two populations of hair bulbs, one with normal iduronate sulphatase activity, reflecting hair bulbs with the normal X chromosome remaining active, and the other with low enzyme activity, representing those with the mutant X chromosome remaining active.

Two populations of hair bulbs with low and normal activity of iduronate sulphatase, respectively, in female carrier of Hunter's syndrome.

DNA analysis is not affected by lyonisation and is the method of choice for detecting carriers. Initial analysis using linked or intragenic probes is being replaced by more direct testing as mutation analysis becomes feasible.

Calculation of the final probability of carrier state often entails analysis of pedigree data with the results of one or more specific tests. The possibility of new mutation and gonadal mosaicism must be taken into account in sporadic cases. The calculation relies on Bayesian analysis, and computer programs are available for the complex analysis required in large families.

Information on consultand:

Prior risk = 50% (mother obligate carrier)

Risk modified by:
  DNA analysis—reducing prior risk to 5%
  One healthy son—reducing risk
  Analysis of serum creatine kinase
    activity—giving probability of
    carrier state of 0·3

Risk after Bayesian calculation = 1%

Calculation of carrier risk in Duchenne muscular dystrophy.

Testing for carrier state

Various methods can be used to determine carrier state; those related directly to gene function discriminate better than those measuring functions further removed from the primary gene defect. Detection of an abnormality confirms the carrier state but its apparent absence does not guarantee normality.

Clinical signs

Careful examination for clinical signs may identify some carriers and is particularly important in autosomal dominant conditions in which the underlying biochemical basis of the disorder is unknown. In some X linked recessive disorders (especially those affecting the eye or skin) abnormalities may be detected in this way in female carriers. The absence of clinical signs does not exclude the carrier state.

Clinical examination can be supplemented with investigations such as physiological studies, microscopy, and radiology. In myotonic dystrophy, for example, before direct mutation analysis became possible asymptomatic carriers could usually be identified in early adult life by a combination of clinical examination to detect myotonia and mild weakness of facial and sternomastoid muscles, slit lamp examination of the eyes to detect lens opacities, and electromyography to look for myotonic changes. Confirmation or exclusion of the carrier state is important for genetic counselling, especially for mildly affected women who have an appreciable risk of producing severely affected infants with the congenital form of myotonic dystrophy.

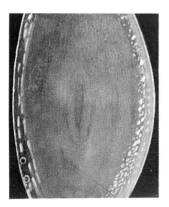

Clinical myotonia, lens opacities, and myotonic discharges on electromyography confirm carrier state in myotonic dystrophy.

17

Detection of carriers

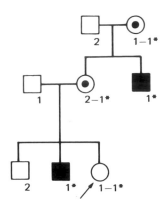

1,2 = DNA variants detected by probe linked to Duchenne muscular dystrophy gene on X chromosome

Prediction of carrier state by DNA analysis in Duchenne muscular dystrophy. Disease gene cosegregates with DNA variant 1*, predicting that consultand (↗) is a carrier.

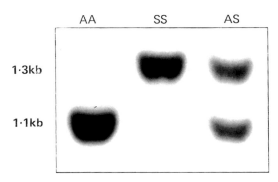

Identification of sickle cell carriers by Southern blot analysis of *Mst II* restoration fragments and β globin gene probe. (AA = Normal, AS = sickle cell trait, SS = sickle cell anaemia.)

Analysis of gene products

Biochemical identification of carriers may be possible when the gene product is known. This approach is used for inborn errors of metabolism due to enzyme deficiency as well as for disorders due to a defective structural protein, such as haemophilia and thalassaemia. Overlap between the ranges of values in heterozygous and normal people occurs even when the primary gene product can be analysed, and interpretation of results can be difficult.

Secondary biochemical abnormalities

When the gene product is not known or cannot be readily tested the identification of carriers may depend on detecting secondary biochemical abnormalities, such as raised serum creatine kinase activity in Duchenne and Becker's muscular dystrophies. The overlap between the ranges of values in normal subjects and carriers is often considerable, and the sensitivity of this type of test is only moderate.

Analysis of genes

DNA analysis has revolutionised predictive testing for genetic disorders. The genes for most important mendelian disorders are now mapped and many have been cloned. Direct mutation analysis is possible for an increasing number of conditions, including the haemoglobinopathies, cystic fibrosis, Duchenne muscular dystrophy, Huntington's disease, myotonic dystrophy, and fragile X syndrome. This provides definitive results for carrier tests, presymptomatic diagnoses, and prenatal diagnoses. Methods of DNA analysis and its application to genetic disease are discussed in later chapters.

The illustration of lens opacities was reproduced by kind permission of Professor P Harper, Institute of Medical Genetics for Wales, Cardiff.

SPECIAL ISSUES

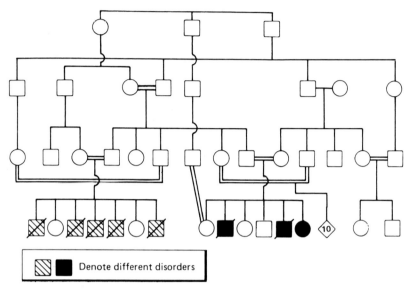

Various issues that arise during genetic counselling need special consideration and are briefly reviewed in this chapter. Consanguinity or disputed paternity affect the assessment of genetic risk, and families with genetic disorders need to know about adoption and other reproductive options. In some genetic conditions population screening may be appropriate in the prevention or early detection of disease. The overall approach to genetic disorders raises various important ethical issues, which need to be faced by the family, the medical profession, and society.

⬛ / ▨	Denote different disorders

Autosomal recessive disorders in a family with complex consanguinity.

Consanguinity

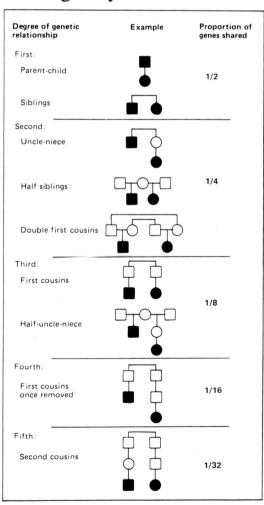

Degree of genetic relationship	Example	Proportion of genes shared
First:		
Parent-child		1/2
Siblings		
Second:		
Uncle-niece		
Half siblings		1/4
Double first cousins		
Third:		
First cousins		1/8
Half-uncle-niece		
Fourth:		
First cousins once removed		1/16
Fifth:		
Second cousins		1/32

Consanguinity is a special problem in genetic counselling because of the increased risk of autosomal recessive disorders. Everyone probably carries at least one harmful autosomal recessive gene. In marriages between first cousins the chance of a child inheriting the same recessive gene from both parents that originated from one of the common grandparents and was transmitted through both sides of the family is one in 64. A different recessive gene may be similarly transmitted from the other common grandparent so that the risk of homozygosity for a recessive disorder in the child is one in 32. If two lethal genes are carried by each person the risk is one in 16.

Marriage between first cousins generally increases the risk of severe abnormality and mortality in offspring, by 3–5% compared with that in the general population. The increased risk associated with marriages between second cousins is around 1%.

Marriage between first and second degree relatives is almost universally illegal, although marriages between uncles and nieces occur in some Asian countries. Marriage between third degree relatives (between cousins or half uncles and nieces) is more common and permitted by law in many countries.

The offspring of incestuous relationships are at high risk of severe abnormality, mental retardation, and childhood death. Only about half of the children born to first degree relatives are normal, and this has important implications for termination of pregnancy or subsequent adoption.

Paternity

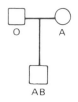

Non-paternity identified by blood group antigens.

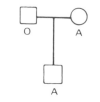

Paternity not excluded by blood group antigens.

Clinical geneticists are sometimes asked to investigate disputed or uncertain paternity, particularly now that parentage can be accurately confirmed or excluded by DNA fingerprinting tests, which remove the uncertainty previously associated with analysis of blood groups. Testing of paternity, however, is not a clinical service, and disputed paternity remains a strictly legal issue.

Non-paternity may be discovered coincidentally during DNA testing of a family to investigate a mendelian disorder. Though this information must remain strictly confidential, it may substantially alter the risks to certain members of the family and is therefore of great importance in subsequent counselling.

Reproductive options

Pregnancy
- With or without prenatal diagnosis

Insemination by donor
- Existing child with autosomal recessive disorder
- Husband has autosomal dominant disorder
- Husband has chromosomal abnormality leading to infertility or recurrent spontaneous abortion

Ovum donation
- Wife has autosomal dominant or X linked disorder
- Wife has chromosomal abnormality leading to infertility or recurrent spontaneous abortion

Contraception
- Couples waiting for new medical developments

Sterilisation
- Couples whose family is complete

Various reproductive options are available to couples for whom pregnancy carries a high risk of abnormalities (box). The acceptability of any of these possibilities is a personal decision of the couple.

Adoption

A couple at risk of transmitting a genetic disorder may wish to consider adopting children as an alternative to pregnancy. The reduction in the availability of babies and young children for adoption should be realised, and, unfortunately, the presence of a genetic disorder in one of the couple may make this option more difficult to achieve. A rigorous assessment is made of prospective adoptive parents, and application takes at least a year. The chances of successful adoption are greater for couples able to accept older children or children with identified problems or handicaps.

A child placed for adoption may have a family history of a genetically determined disease, such as schizophrenia, and this may indeed be the reason for adoption. Some serious genetic disorders can be identified in infancy, others may not become apparent until adult life. Assessing the risk to the child before adoption is important, and adoptive parents should be given appropriate information about the risks and their implications. Generally, children with confirmed genetic disorders or those at risk should not be considered to be unsuitable for adoption as many adoptive parents elect to proceed with adoption after the possibilities for the child's future have been discussed.

Screening for genetic disorders

Screening pregnancies at risk
Neural tube defect

Down's syndrome

Neonatal screening
Phenylketonuria

Hypothyroidism

Screening for carriers
Thalassaemia

Sickle cell disease

Tay-Sachs diseases

Cystic fibrosis

Screening programmes may be designed to diagnose genetic disorders or to identify couples at risk of transmitting genetic disorders to their children. Screening tests must be sufficiently sensitive to avoid false negative results and yet specific enough to avoid false positive results. To be employed on a large scale the tests must also be safe, simple, and fairly inexpensive. Screening programmes need to confer benefits to individual subjects as well as to society and to be successful stigmatisation must be avoided.

Screening during pregnancy for neural tube defect, based on serum screening or ultrasound scanning, or both, is offered by all obstetric centres. Estimation of maternal serum α fetoprotein concentration identifies over 90% of fetuses with anencephaly and around 80% of those with open neural tube defect. In experienced hands ultrasound scanning detects almost all babies with neural tube defect.

Child of normal intelligence
treated for phenylketonuria.

Collecting a mouthwash sample for DNA
extraction and carrier testing for cystic fibrosis.

Screening for Down's syndrome has previously been based on maternal age, which does not identify all cases. A detection rate of 35% could be achieved if all women aged over 35 had amniocentesis during their pregnancy. The rate of detection can be improved to over 60% by incorporating the results of measurements of serum α fetoprotein, unconjugated oestriol, and human chorionic gonadotrophin concentrations with maternal age to give a composite risk value.

There are well established programmes for screening all neonates for phenylketonuria and hypothyroidism, and early diagnosis and treatment is successful in preventing mental retardation in the affected children. The value of including other metabolic disorders in a screening programme would depend on the incidence of the disorder and the prospect of altering the prognosis by its early detection. Possible candidates include galactosaemia, maple syrup urine disease, and congenital adrenal hyperplasia.

Population screening aimed at identifying carriers of common autosomal recessive disorders allows the identification of carrier couples before they have an affected child, and provides the opportunity for prenatal diagnosis. Carrier screening programmes for thalassaemia and Tay-Sachs disease in high risk ethnic groupings in several countries have resulted in a significant reduction in the birth prevalence of these disorders. Carrier screening for cystic fibrosis is now possible, although not all carriers can be identified because of the diversity of mutations within the cystic fibrosis gene. Screening programmes instituted in antenatal clinics and in general practice report a substantial uptake for carrier testing. It is important that appropriate information and counselling is available to individuals being offered screening, as they are likely to have little prior knowledge of the disorder being screened for and the implications of a positive test. Training in this area will be needed by members of primary health care teams as screening will probably occur largely within a general practice setting.

Ethical issues

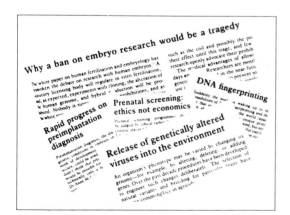

Many ethical controversies occur in clinical genetics. Predictive genetic testing, embryo research, and potential gene therapy are some that are currently debated. New technologies often generate concern over their potential application and safety. Widely publicised fears about the possible danger of genetic engineering accidents have proved unfounded, and recombinant DNA technology now plays an important part in investigating genetic disease.

In clinical practice the preservation of patient confidentiality may conflict with the need to disclose particular information to other family members. If a patient refuses to allow information about himself or herself to be disclosed the doctor may have to break confidentiality to inform immediate relatives of their own risk of developing or transmitting a disorder. Moral dilemmas also face many families with genetic disorders—for example, couples may have to make decisions about prenatal diagnosis and termination of pregnancy. Personal convictions about such issues vary widely, and it is important that couples should be allowed to make their own decisions with the help of non-directive medical information.

CHROMOSOMAL ANALYSIS

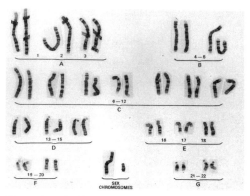

Normal male chromosome constitution.

The correct chromosome complement in humans was established in 1956, and the first chromosomal disorders (Down's, Turner's, and Klinefelter's syndromes) were defined in 1959. Since then refinements in techniques of preparing and examining samples have led to the description of hundreds of disorders that are due to chromosomal abnormalities.

Description of terms

Euploid	Chromosome numbers are multiples of the haploid set (2n)
Polyploid	Chromosome numbers are greater than diploid (3n, triploid)
Aneuploid	Chromosome numbers are not exact multiples of the haploid set (2n+1 trisomy; 2n−1 monosomy)
Mosaic	Presence of two different cell lines derived from one zygote (46XX/45X, Turner's mosaic)
Chimaera	Presence of two different cell lines derived from fusion of two zygotes (46XX/46XY, true hermaphrodite)

Human somatic cells contain 46 chromosomes organised into 22 autosomal pairs plus sex chromosomes. The basic haploid set ($n = 23$) is present in the gametes. After fertilisation the zygote contains a diploid set of chromosomes ($2n = 46$); one of each pair is maternal in origin, the other paternal.

During meiosis, which is the nuclear division giving rise to the gametes, recombination occurs between homologous parental chromosomes. The exchange of chromosomal material leads to the separation of genes originally located on the same chromosome, and gives rise to genetic variation within families.

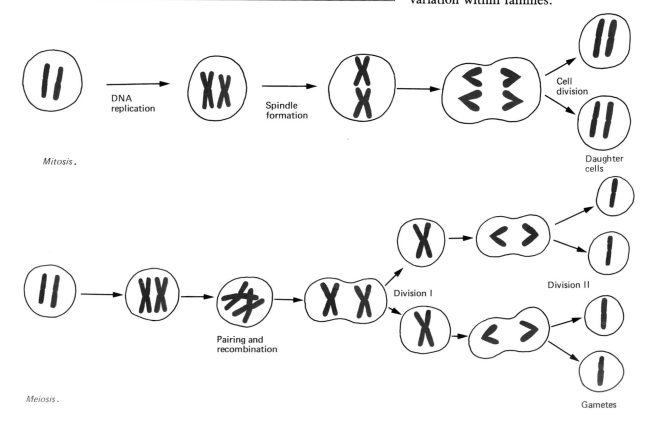

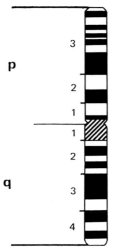

Simplified banding pattern of chromosome 1.

Each chromosome can be identified by light microscopy with staining techniques that give a characteristic pattern of alternating light and dark bands. During metaphase the two chromatids of each chromosome are joined at the centromere. The short arm of the chromosome is designated p and the long arm q. Each arm is subdivided numerically into a number of bands, according to the Paris convention, which permits precise localisation of a structural abnormality. High resolution cytogenetic techniques have permitted identification of small interstitial chromosome deletions in recognised disorders of previously unknown origin, such as Prader-Willi and Angelman's syndromes. Deletions too small to be detected by microscopy may be amenable to diagnosis by DNA techniques.

Types of chromosomal disorders

Type of disorder	Example		Outcome
Numerical			
Polyploid	Triploidy	69 chromosomes	Lethal
Aneuploid	Trisomy of chromosome 21		Down's syndrome
	Monosomy of X chromosome		Turner's syndrome
	47 chromosomes (XXY)		Klinefelter's syndrome
Structural			
Deletion	Terminal deletion 5p		Cri du chat syndrome
	Interstitial deletion 11p		Found in Wilms's tumour
Inversion	Pericentric inversion 9		Normal phenotype
Duplication	Isochromosome X (fusion of long arms with loss of short arms)		Infertility in females
Ring chromosome	Ring chromosome 18		Mental retardation syndrome
Fragile site	Fragile X		Mental retardation syndrome
Translocation	Reciprocal		Balanced translocations cause no abnormality. Unbalanced translocations cause spontaneous abortions or syndromes of multiple physical and mental handicaps
	Robertsonian		

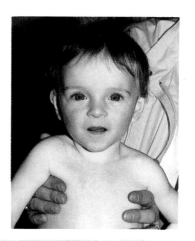

Cri du chat syndrome associated with deletion of short arm of chromosome 5.

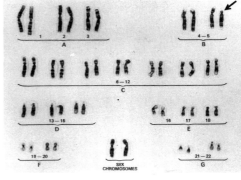

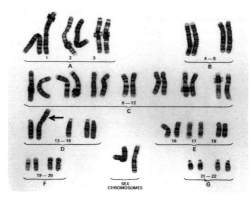

Balanced Robertsonian translocation affecting chromosomes 13 and 14.

Chromosomal analysis

Reporting of karyotypes

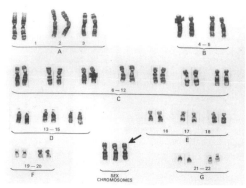

47, XXX karyotype in triple X syndrome.

- Total number of chromosomes given first followed by constitution of sex chromosomes:

46, XX	Normal female
47, XXY	Male with Klinefelter's syndrome
47, XXX	Female with triple X syndrome

- Additional or lost chromosomes are indicated by + or −:

47, XY, +21	Male with trisomy 21 (Down's syndrome)
46, XX, 12p+	Additional unidentified material on short arm of chromosome 12

- All cell lines present are shown for mosaics:

46, XX/47, XX, +21	Down's mosaic
46, XX/47, XXX/45, X	Turner's/triple X mosaic

- Structural rearrangements are described, identifying p and q arms and location of abnormality:

46, XY, del 11 (p13)	Deletion of short arm of chromosome 11 at band 13
46, XX, t (X;7) (p21;q23)	Translocation between chromosome X and 7 with break points in respective chromosomes

Molecular cytogenetics

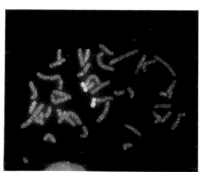

Fluorescence in situ hybridisation of normal metaphase chromosomes hybridised with chromosome 20 probes derived from the whole chromosome, which identify each individual chromosome 20.

Fluorescence in situ hybridisation (FISH) is a recently developed molecular cytogenetic technique, involving hybridisation of a DNA probe to a metaphase chromosome spread. Single stranded probe DNA is fluorescently labelled using biotin and avidin and hybridised to the denatured DNA of intact chromosomes on a microscope slide. The resultant DNA binding can be seen direct using a fluorescence microscope.

One application of FISH is in chromosome painting. This technique uses an array of specific DNA probes derived from a whole chromosome and causes the entire chromosome to fluoresce. This can be used to identify the chromosomal origin of structural rearrangements that cannot be defined by conventional cytogenetic techniques.

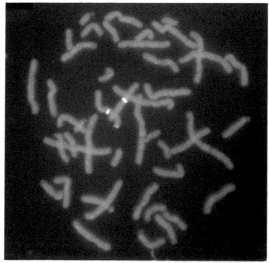

Fluorescence in situ hybridisation of metaphase chromosomes from a male with 46 XX chromosome constitution hybridised with separate probes derived from both X and Y chromosomes. The X chromosome probe (yellow) has hybridised to both X chromosomes. The Y chromosome probe (red) has hybridised to one of the X chromosomes, which indicates that this chromosome carries Y chromosomal DNA, thus accounting for the subject's phenotypic sex.

Alternatively, a single DNA probe corresponding to a specific locus can be used. Hybridisation reveals fluorescent spots on each chromatid of the relative chromosome. This method is used to detect the presence or absence of specific DNA sequences and is useful in the diagnosis of syndromes caused by sub-microscopic deletions, or in identifying carriers of single gene defects due to large deletions, such as Duchenne muscular dystrophy.

It is possible to use several separate DNA probes, each labelled with a different fluorochrome, to analyse more than one locus or chromosome region in the same reaction. Another application of this technique is in the study of interface nuclei, which permits the study of non-dividing cells. Thus, rapid results can be obtained for the diagnosis or exclusion of Down's syndrome in uncultured amniotic fluid samples using chromosome 21 specific probes.

Incidence of chromosomal abnormalities

Incidence of chromosomal abnormalities in spontaneous abortions and stillbirths	%
Spontaneous abortions:	
All	50
Before 12 weeks	60
12–20 Weeks	20
Stillbirths	5

Types of chromosomal abnormalities in spontaneous abortions	%
Trisomy	52
Monosomy X	18
Triploidy	17
Translocations	2–4

Chromosomal abnormalities in newborn infants (per 1000)	
All	6·5
Autosomal trisomy	1·7
Autosomal rearrangements	1·9
Other autosomal abnormality	0·4
Sex chromosomal	2·5

Common abnormalities

Autosomal
Trisomy 21—Down's ⎫
Trisomy 18—Edwards's ⎬ syndrome
Trisomy 13—Patau's ⎭

Sex chromosomal
XO—Turner's ⎫
XXX—Triple X ⎬ syndrome
XXY—Klinefelter's ⎭
XYY—XYY Male

Chromosomal abnormalities are particularly common in spontaneous abortions. About 15–20% of all conceptions are estimated to be lost spontaneously, and about half of these are associated with a chromosomal abnormality. Most chromosomal abnormalities lead to spontaneous abortion, some inevitably so—for example, trisomy 16 is commonly found in aborted fetuses but never in liveborn infants.

In liveborn infants chromosomal abnormalities occur at about six per 1000 births. The incidence of abnormalities of autosomes and sex chromosomes is about the same. The effect on the child depends on the type of abnormality. Abnormalities do not occur in balanced rearrangements and are mild in disorders of the sex chromosomes. Unbalanced autosomal abnormalities cause disorders with multiple congenital malformations, almost invariably associated with mental retardation.

The illustrations of normal male, cri du chat, Robertsonian translocation 13; 14, triple X karyotypes, and fluorescence in situ hybridisation were reproduced by kind permission of Dr Lorraine Gaunt, St Mary's Hospital, Manchester.

COMMON CHROMOSOMAL DISORDERS

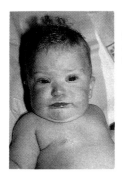

Child with developmental delay and deletion of chromosome 13.

Chromosomal abnormalities generally cause multiple congenital malformations and mental retardation. Children with more than one physical abnormality, particularly if retarded, should therefore undergo chromosomal analysis as part of their investigation. Chromosomal disorders are incurable but can be reliably detected by prenatal diagnostic techniques. Amniocentesis or chorionic villus sampling should be offered to women whose pregnancies are at increased risk—namely, women identified by biochemical screening for Down's syndrome, couples with an affected child, and couples in whom one partner carries a balanced translocation. Unfortunately, when there is no history of previous abnormality the risk in many affected pregnancies cannot be predicted before the child is born.

Autosomal abnormalities

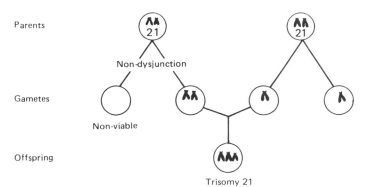

Non-dysjunction of chromosome 21 leading to Down's syndrome.

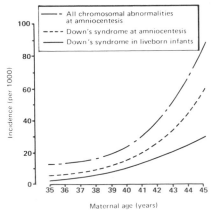

Incidence of chromosomal abnormalities and Down's syndrome by maternal age.

Risk for trisomy 21 in liveborn infants by maternal age

Maternal age at delivery	Risk
All ages	1 in 650
Age 30	1 in 900
Age 35	1 in 400
Age 36	1 in 300
Age 37	1 in 250
Age 38	1 in 200
Age 39	1 in 150
Age 40	1 in 100
Age 44	1 in 40

Trisomy 21 (Down's syndrome)

Down's syndrome is the commonest autosomal trisomy, the incidence in liveborn infants being one in 650, although more than half of conceptions with trisomy 21 do not survive to term. Affected children have a characteristic facial appearance, are mentally retarded, and may die young. They may have associated congenital heart disease and are at increased risk for recurrent infections, atlantoaxial instability, and acute leukaemia.

Most cases are due to non-dysjunction of chromosome 21 during meiosis in the formation of eggs or sperm. Although occurring at any age, non-dysjunction increases with maternal age. The risk of recurrence for a chromosomal abnormality in a liveborn infant after the birth of a child with trisomy 21 is about 1% (0·5% for trisomy 21 and 0·5% for other chromosomal abnormalities). For mothers aged 35 and over the total risk is around four times the risk related to age quoted in the table, half being for Down's syndrome and half for other chromosomal abnormalities.

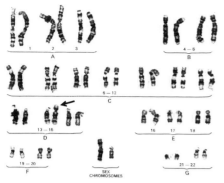

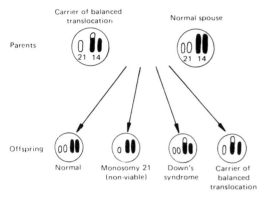

Down's syndrome due to Robertsonian translocation between chromosomes 14 and 21.

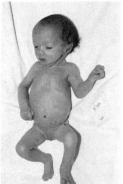

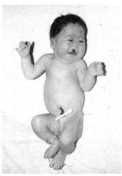

Trisomy 21 cell line in mosaic Down's syndrome. (Normal cell line also present.)

Trisomy 18—skull shape and facial features, short sternum, clenched hands, and rocker-bottom feet.

Trisomy 13—facial appearance associated with holoprosencephaly; postaxial polydactyly of hands and feet.

Possible chromosome arrangements in offspring of a carrier of a balanced 14;21 translocation

Girl with mosaic trisomy 21

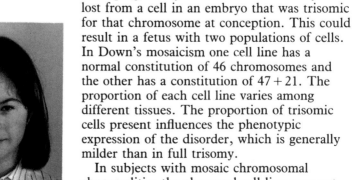

About 5% of cases of Down's syndrome are due to translocation, in which chromosome 21 is translocated on to chromosome 14 or, occasionally, chromosome 22. In less than half of these cases one of the parents has a balanced version of the same translocation. A healthy adult with a balanced translocation has 45 chromosomes, and the affected child has 46 chromosomes, the extra chromosome 21 being present in the translocation form.

The risk of Down's syndrome in the offspring is 10% when the balanced translocation is carried by the mother and 2·5% when carried by the father. If neither parent has a balanced translocation, an affected child represents a spontaneous, newly arising event, and the risk of recurrence is low (<1%).

Occasionally, Down's syndrome is due to a 21;21 translocation. A parent with a balanced translocation would be unable to have normal children.

When a case of translocation occurs it is important to test other family members to identify all carriers of the translocation whose pregnancies would be at risk.

Couples concerned about a family history of Down's syndrome can have their chromosomes analysed from a sample of blood to exclude a balanced translocation if the karyotype of the affected person is not known. This usually avoids unnecessary amniocentesis during pregnancy.

Trisomy 18 (Edwards's syndrome)

Trisomy 18 has an overall incidence of around 0·12 per 1000 live births. As with Down's syndrome most cases are due to non-dysjunction and the incidence increases with maternal age. Risk of recurrence is low, unless due to a parental translocation. Affected infants usually succumb within a few weeks or months but may occasionally survive several years. The main features include mental deficiency, growth deficiency, characteristic facial appearance, clenched hands, rocker bottom feet, and cardiac and renal abnormalities.

Trisomy 13 (Patau's syndrome)

The incidence of trisomy 13 is about 0·07 per 1000 live births, mainly due to non-dysjunction, with low risk of recurrence. Cases of translocation with higher risk of recurrence also occur. Most affected infants succumb within hours or weeks of birth. The main features include severe mental deficit; structural abnormalities of the brain, including microcephaly and holoprosencephaly (a developmental defect of the forebrain); cleft lip and palate; polydactyly; and ophthalmic, cardiac, and renal malformations.

Mosaics

After fertilisation of a normal egg non-dysjunction may occur during a mitotic division in the developing embryo or a chromosome may be lost from a cell in an embryo that was trisomic for that chromosome at conception. This could result in a fetus with two populations of cells. In Down's mosaicism one cell line has a normal constitution of 46 chromosomes and the other has a constitution of 47 + 21. The proportion of each cell line varies among different tissues. The proportion of trisomic cells present influences the phenotypic expression of the disorder, which is generally milder than in full trisomy.

In subjects with mosaic chromosomal abnormalities the abnormal cell line may not be present in peripheral lymphocytes, and skin biopsy and culturing of the cells is often required for diagnosis.

Common chromosomal disorders

Normal 8 month old infant born after trisomy 20 mosaicism detected in amniotic cells.

The clinical importance of a mosaic abnormality that is detected by amniocentesis can be difficult to interpret. Mosaicism of chromosome 20, for example, is not usually associated with fetal abnormality. Mosaicism for a marker (small unidentified) chromosome carries a much smaller risk of causing mental retardation if familial, and therefore the parents need to be investigated before advice can be given. Mosaicism detected in chorionic villus samples may reflect an abnormality confined to placental tissue that does not affect the fetus.

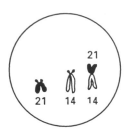

Balanced Robertsonian translocation affecting chromosomes 14 and 21.

Unbalanced Robertsonian translocation affecting chromosomes 14 and 21 and resulting in Down's syndrome.

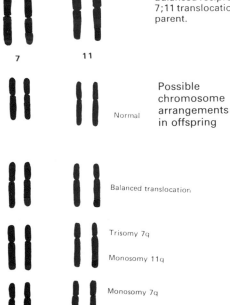

Balanced reciprocal 7;11 translocation in parent.

7   11

Normal

Possible chromosome arrangements in offspring

Balanced translocation

Trisomy 7q

Monosomy 11q

Monosomy 7q

Trisomy 11q

Translocations

Translocations may be of two types: Robertsonian or reciprocal. Robertsonian translocations occur when two of the acrocentric chromosomes (13, 14, 15, 21, or 22) become joined together. Balanced translocation carriers have 45 chromosomes but no significant loss of overall chromosomal material and are almost always healthy. In unbalanced translocation karyotypes there are 46 chromosomes with trisomy for one of the chromosomes involved in the translocation. This may lead to spontaneous miscarriage (chromosomes 14, 15, and 22) or liveborn infants with trisomy (chromosomes 13 and 21). Unbalanced Robertsonian translocations may arise spontaneously or be inherited from a parent carrying a balanced translocation.

Reciprocal translocations involve exchange of chromosomal segments between two different chromosomes, generated by chromosomal breakage and rejoining. Balanced translocations generally occur in healthy people, but there is some risk of mental retardation in subjects with apparently balanced de novo translocations due to loss of DNA that cannot be visualised on routine microscopic chromosomal analysis.

Abnormalities resulting from an unbalanced translocation karyotype depend on the particular chromosome fragments that are present in monosomic or trisomic form. Sometimes spontaneous abortion is inevitable; at other times a child with multiple abnormalities may be born alive. The risk of an unbalanced karyotype occurring in offspring depends on the individual translocation.

Once a translocation has been identified it is important to investigate relatives of that person to identify carriers of the balanced translocation whose offspring would be at risk. Pregnancies can be monitored with chorionic villus sampling or amniocentesis.

Microdeletions

Several genetic syndromes have now been identified by molecular cytogenetic techniques as being due to chromosomal deletions too small to be seen by conventional analysis. These are termed submicroscopic deletions or microdeletions and probably affect less than 4000 kilobases of DNA. A microdeletion may involve a single gene, or extend over several genes. The term continuous gene syndrome is applied when several genes are affected, and in these disorders the features present may be determined by the extent of the deletion. The chromosomal location of a microdeletion may be identified by the presence of a larger visible cytogenetic deletion in a proportion of cases, as in Prader-Willi and Angelman's syndrome, or by finding a chromosomal translocation in an affected individual, as occured in William's syndrome.

Typical facial appearance of William's syndrome in child with supravalvular aortic stenosis and delayed development.

A microdeletion on chromosome 22q11 has been found in most cases of DiGeorge's syndrome and velocardiofacial syndrome, and is also associated with certain types of isolated congenital heart disease. With an incidence of 8 per 1000 live births, congenital heart disease is one of the most common malformations. The aetiology is usually unknown and it is therefore important to identify cases caused by 22q11 deletion.

DiGeorge's syndrome involves thymic aplasia, parathyroid hypoplasia, aortic arch and conotruncal anomalies, and characteristic facies due to defects of 3rd and 4th branchial arch development. Velocardiofacial syndrome was described as a separate clinical entity, but does share many features in common with DiGeorge's syndrome. The features include mild mental retardation, short stature, cleft palate or speech defect due to palatal dysfunction, prominent nose, and congenital cardiac defects including ventricular septal defect, right sided aortic arch, and tetralogy of Fallot.

Examples of syndromes associated with microdeletions	
Syndrome	Chromosomal deletion
DiGeorge's	22q11
Velocardiofacial	22q11
Prader-Willi	15q11-13
Angelman's	15q11-13
William's	7q11
Miller-Dieker (lissencephaly)	17p13
Rubinstein-Taybi	16

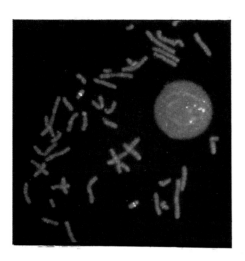

Fluorescence in situ hybridisation with a probe from the DiGeorge's critical region of chromosome 22q11, which shows normal hybridisation to both chromosome 22s and these are therefore not deleted in this region.

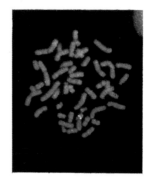

Fluorescence in situ hybridisation with a probe from the DiGeorge's critical region of chromosome 22q11, which shows hybridisation to only one chromosome 22, thus indicating that the other chromosome 22 is deleted in this region.

Isolated cardiac defects due to microdeletions of chromosome 22q11 often include outflow tract abnormalities. Deletions have been observed in both sporadic and familial cases and are responsible for about 30% of non-syndromic conotruncal malformations including interrupted aortic arch, truncus arteriosus, and tetralogy of Fallot.

Sex chromosomal abnormalities

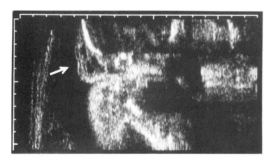

Cystic hygroma in Turner's syndrome detected by ultrasonography.

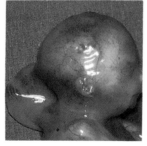

Fetus with Turner's syndrome.

Numerical abnormalities of the sex chromosomes are fairly common and cause less severe defects than autosomal abnormalities. They are often detected coincidentally at amniocentesis or during investigation for infertility, and risk of recurrence in families is low. When more than one additional sex chromosome is present mental retardation or physical abnormality is more likely.

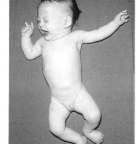

Lymphoedema of the feet as only manifestation of Turner's syndrome in newborn infant.

Turner's syndrome

Turner's syndrome results in early spontaneous loss of the fetus in over 95% of cases. Severely affected fetuses who survive to the second trimester can be detected by ultrasonography, which shows cystic hygroma, chylothorax, asictes, and hydrops.

The incidence of Turner's syndrome in liveborn female infants is 0·4 per 1000. Phenotypic abnormalities vary considerably but are usually mild. In some infants the only detectable abnormality is lymphoedema of the hands and feet. The most consistent features of the syndrome are short stature and infertility, but neck webbing, cubitus valgus, and aortic coarctation may also occur. Intelligence is usually within the normal range, but a few girls have educational problems. Growth can be stimulated with androgens or growth hormone, and oestrogen replacement treatment is necessary for pubertal development.

Common chromosomal disorders

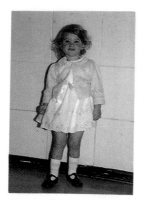

Normal appearance and development in 22 month girl with triple X syndrome.

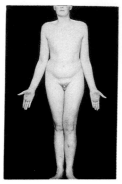

Tall stature, truncal obesity, and underdeveloped genitalia in Klinefelter's syndrome.

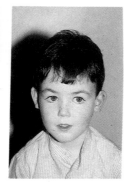

Normal facial appearance in mildly retarded boy with 48, XYYY karyotype.

Triple X syndrome

The triple X syndrome occurs with an incidence of 0·65 per 1000 liveborn female infants and is usually a coincidental finding. Apart from being taller than average, affected girls are physically normal. Educational problems are encountered more often in this group than in the other types of sex chromosomal abnormalities. Mean intelligence quotient is lower than in controls, about half of affected girls having delayed speech development and three quarters requiring some remedial teaching. Gonadal function is usually normal, but premature ovarian failure may occur.

Klinefelter's syndrome

The XXY karyotype of Klinefelter's syndrome occurs with an incidence of 2·0 per 1000 liveborn males. The primary feature of the syndrome is hypogonadism, and affected males are usually tall. Pubertal development often progresses normally, but testosterone replacement treatment is sometimes required. Testicular size decreases after puberty, and affected males are infertile. Gynaecomastia may occur, and the risk of cancer of the breast is increased. Intelligence is generally within the normal range, but educational difficultes and behavioural problems are fairly common.

XYY syndrome

The XYY syndrome occurs in about 1·5 per 1000 liveborn male infants. Although more prevalent among inmates of high security institutions, the syndrome is less strongly associated with aggressive behaviour than previously thought, and many affected males remain undetected clinically. Mild mental retardation and behavioural problems can occur, and tall stature is usual.

Illustrations reproduced by kind permission of colleagues at St Mary's Hospital, Manchester, were: Down's syndrome karyotype and fluorescence in situ hybridisation, Dr Lorraine Gaunt; cystic hygroma scan, Dr Sylvia Rimmer; trisomy 13, trisomy 18, Turner's syndrome fetus, and Klinefelter's syndrome, Professor Dian Donnai.

GENETICS OF COMMON DISORDERS

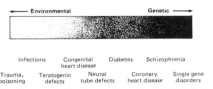

Relative contribution of environmental and genetic factors in some common disorders.

Multifactorial inheritance

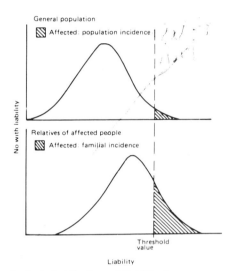

Hypothetical distribution of liability for a multifactorial disorder in general population and affected families.

Empirical recurrence risks to siblings in Hirschsprung's disease, according to sex of person affected and length of aganlionic segment

Length of colon affected	Sex of person affected	Risk to siblings (%)	
		Brothers	Sisters
Short segment	Male	4·7	0·6
	Female	8·1	2·9
Long segment	Male	16·1	11·1
	Female	18·2	9·1

Risk of recurrence

Factors increasing risk to relatives in multifactorial disorders

- High heritability of disorder
- Close relationship to proband
- Multiple affected family members
- Severe disease in proband
- Proband being of sex not usually affected

The genetic contribution to disease varies; some disorders are entirely environmental and others are wholly genetic. Many common disorders, however, have an appreciable genetic contribution but do not follow simple patterns of inheritance within a family. The terms multifactorial or polygenic inheritance have been used to describe the aetiology of these disorders. Normal traits inherited in this way include height and intelligence.

The concept of multifactorial inheritance implies that a disease is caused by the interaction of several adverse genetic and environmental factors. The liability of a population to a particular disease follows a normal distribution curve, most people showing only moderate susceptibility and remaining unaffected. Only when a certain threshold of liability is exceeded is the disorder manifest. Relatives of an affected person will show a shift in liability, with a greater proportion of them being beyond the threshold. Familial clustering of a particular disorder may therefore occur.

Unravelling the molecular genetics of the complex multifactorial diseases is much more difficult than for single gene disorders. Nevertheless, this is an important task as these diseases account for the great majority of morbidity and mortality in developed countries. Approaches to multifactorial disorders include the identification of disease associations in the general population, linkage analysis in affected families, and the study of animal models. Identification of genes causing the familial cases of diseases that are usually sporadic, such as Alzheimer's disease and motor neurone disease, may give insights into the pathogenesis of the more common sporadic forms of the disease. In the future, understanding genetic susceptibility may allow for screening for, and prevention of, common diseases.

Genetic susceptibility to common disorders may be due to a limited, rather than a large, number of loci. Several common disorders thought to follow polygenic inheritance (such as diabetes, hypertension, congenital heart disease, and Hirschsprung's disease) have been found in some individuals and families to be due to single gene defects. In Hirschprung's disease (aganglionic megacolon) family data on recurrence risks support the concept of sex-modified polygenic inheritance, although autosomal dominant inheritance with reduced penetrance has been suggested in some families with several affected members. Mutations in the ret proto-oncogene on chromosome 10q11·2 or in the endothelin-B receptor gene on chromosome 13q22 have been detected in both familial and sporadic cases, indicating that a proportion of cases are due to a single gene defect.

The risk of recurrence for a multifactorial disorder within a family is generally low and mainly affects first degree relatives. In many conditions family studies have reported the rate with which relatives of the proband have been affected. This allows empirical values for risk of recurrence to be calculated, which can be used in genetic counselling. A rational approach to preventing the disease is to modify known environmental triggers in genetically susceptible subjects. Folic acid supplementation in pregnancies at increased risk of neural tube defect and modifying diet and smoking habits in coronary heart disease are examples of effective intervention, but this approach is not currently possible for many disorders.

Heritability

Estimates of heritability

	Heritability (%)
Schizophrenia	85
Asthma	80
Cleft lip and palate	76
Coronary heart disease	65
Hypertension	62
Neural tube defect	60
Peptic ulcer	37

The genetic contribution to the aetiology of a disorder, or heritability, can be calculated from the disease incidence in the general population and that in relatives of an affected subject. Disorders with a greater genetic contribution have higher heritability and, hence, higher risks of recurrence.

HLA association and linkage

Diseases associated with histocompatibility antigens

Ankylosing spondylitis	B27
Autoimmune thyroid disease	B8, DR3
Chronic active hepatitis	B8, DR3
Coeliac disease	B8, DR3
Diabetes (juvenile)	{ B8, DR3 / B15, DR4
Haemochromatosis	A3
Multiple sclerosis	DR2
Psoriasis	CW6
Reiter's disease	B27
Rheumatoid arthritis	DR4

Several important disorders occur more commonly than expected in subjects with particular HLA phenotypes, which implies that certain HLA determinants may affect disease susceptibility. Awareness of such associations may be helpful in counselling. For example, ankylosing spondylitis, which has an overall risk of recurrence of 4% in siblings, shows a strong association with HLA-B27, and 95% of affected people are positive for this antigen. The risk to their first degree relatives is increased to 9% for those who are also positive for HLA-B27 but reduced to less than 1% for those who are negative.

Genetic association, which may imply a causal relation, is different from genetic linkage, which occurs when two gene loci are physically close together on the chromosome. A disease gene, located near the HLA complex of genes on chromosome 6, will be linked to a particular HLA haplotype within a given affected family but will not necessarily be associated with the same HLA antigens in unrelated affected people. HLA typing can be used to predict disease by establishing the linked HLA haplotype within a given family.

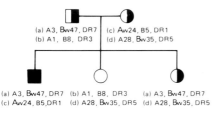

(a) A3, Bw47, DR7   (c) Aw24, B5, DR1
(b) A1, B8, DR3   (d) A28, Bw35, DR5

(a) A3, Bw47, DR7   (b) A1, B8, DR3   (a) A3, Bw47, DR7
(c) Aw24, B5,DR1   (d) A28, Bw35, DR5 (d) A28, Bw35, DR5

■ Homozygous affected

◨ ◖ Heterozygous carrier

Inheritance of congenital adrenal hyperplasia (21-hydroxylase deficiency) and HLA haplotypes (a) and (c).

Congenital adrenal hyperplasia due to 21-hydroxylase deficiency shows both linkage and association with histocompatibility antigens. The 21-hydroxylase gene lies within the HLA gene cluster and is therefore linked to the HLA haplotype. In addition, the salt losing form of 21-hydroxylase deficiency is associated with HLA-Bw47 antigen. This combination of linkage and association is known as linkage disequilibrium and results in certain alleles at neighbouring loci occurring together more often than would be expected by chance.

Twins

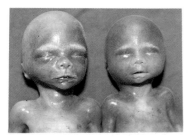

Twin fetuses discordant for Down's syndrome (affected twin on right).

Twins share a common intrauterine environment, but though monozygous twins are genetically identical, dizygous twins are no more alike than any other pair of siblings. This provides the basis for studying twins to determine the genetic contribution in various disorders, by comparing the rates of concordance or discordance for a particular trait between pairs of monozygous and dizygous twins. The rate of concordance in monozygous twins is high for disorders in which genetic predisposition plays a major part in the aetiology of the disease. The phenotypic variability of genetic traits can be studied in monozygous twins, and the effect of a shared intrauterine environment may be studied in dizygous twins.

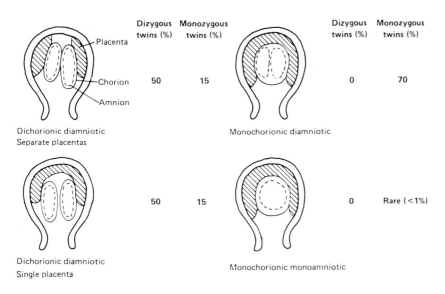

	Dizygous twins (%)	Monozygous twins (%)
	50	15

Dichorionic diamniotic
Separate placentas

	Dizygous twins (%)	Monozygous twins (%)
	0	70

Monochorionic diamniotic

	Dizygous twins (%)	Monozygous twins (%)
	50	15

Dichorionic diamniotic
Single placenta

	Dizygous twins (%)	Monozygous twins (%)
	0	Rare (<1%)

Monochorionic monoamniotic

Twins may be derived from a single egg (monozygous, identical) or two separate eggs (dizygous, fraternal). Examination of the placenta and membranes may help to distinguish between monozygous and dizygous twins but is not completely reliable. Monozygosity—resulting in twins of the same sex who look alike—can be confirmed by investigating inherited characteristics such as blood group markers or DNA polymorphisms (fingerprinting).

Dizygous twins may be familial and are more common in black people than in white Europeans. Monozygous twins are seldom familial and have a fairly constant incidence of 0·4% of pregnancies. Monozygous twin pregnancies are associated with twice the risk of congenital malformation as singleton or dizygous twin pregnancies.

Diabetes

General distinction between insulin dependent and non-insulin dependent diabetes

	Insulin dependent diabetes	Non-insulin dependent diabetes
Clinical features	Thinness / Ketosis / Early onset	Obesity / No ketosis / Late onset
Treatment	Insulin	Diet or drugs
Concordance in monozygotic twins	50%	100%
Histocompatibility antigens	Associated	Not associated
Autoimmune disease	Associated	Not associated
Antibodies to insulin and islet cells	Present	Absent

A genetic predisposition is well recognised in both type I insulin dependent diabetes and type II non-insulin dependent diabetes. Maturity onset diabetes of the young is a specific form of non-insulin dependent diabetes that follows autosomal dominant inheritance. Clinical diabetes or impaired glucose tolerance also occurs in several genetic syndromes—for example, haemochromatosis, growth hormone deficiency, and Wolfram syndrome (diabetes mellitus, optic atrophy, diabetes insipidus, and deafness). Only rarely is diabetes caused by the secretion of an abnormal insulin molecule.

Type I (insulin dependent) diabetes is heterogeneous, and the genetics of the disease are not well understood. The disorder affects about 3 per 1000 of the population in the UK and is a T cell dependent autoimmune disease. Genetic predisposition is important, but less then 50% of monozygous twins are concordant for the disease. The risk to siblings is about 6%, rising to 12% for HLA identical siblings, and this indicates that environmental factors (such as triggering viral infections) are also involved. About 60–70% of the genetic susceptibility is likely to be HLA encoded, with 30–40% attributable to genes outside the HLA region. An association with DR3 and DR4 class II antigens is well documented, 95% of people with insulin dependent diabetes having one or both antigens, compared with 50–60% of the normal population. As most people with DR3 or DR4 class II antigens do not develop diabetes, these antigens are unlikely to be the primary susceptibility determinants. Better definition of susceptible genotypes is becoming possible as subgroups of DR3 and DR4 serotypes are defined by molecular analysis. Evidence suggests involvement of other class II antigens encoded by specific alleles of the DQB1 genes in conferring DR3 and DR4 associated susceptibility.

Factors indicating increased risk of insulin dependent diabetes

- HLA haplotypes shared with affected sibling
- DR3/DR4 antigens
- Insulin autoantibodies
- Islet cell antibodies
- Activated T lymphocytes

Genetics of common disorders

Empirical risk for diabetes according to affected members of family

	Risk (%)
Insulin dependent diabetes:	
Sibling	3–10
One parent	3
Both parents	20
Monozygous twin	50
Non-insulin dependent diabetes:	
First degree relative	10–40
Monozygous twin	100
Maturity onset diabetes of the young:	
First degree relative	50

Certain HLA serotypes, notably DRW6 and DR2 and DR7, confer protection against diabetes, and these may again be associated with particular DQB1 and DQA1 alleles. The role of non-HLA genes in conferring susceptibility to type 1 diabetes is less well understood.

Non-insulin dependent diabetes mellitus shows a strong genetic predisposition, and concordance in monozygotic twins is almost 100%. The risk to siblings may approach 40% by the age of 80, but this high risk is for a disorder that is generally mild. The genetics of this type of diabetes are not understood, and genetically susceptible people cannot be identified with certainty, although certain alleles in the hypervariable region near the insulin gene detected by DNA analysis are found more commonly in people with non-insulin dependent diabetes than in the general population. Other factors, such as obesity, are also implicated in the aetiology.

Coronary heart disease

Environmental contributions to coronary heart disease

- Smoking
- Overweight
- High blood pressure

Environmental factors play a considerable part in coronary heart disease and many risk factors have been identified including raised plasma lipids, impaired glucose tolerance, raised blood pressure, and body mass index and cigarette consumption. There is also an underlying genetic susceptibility, as the risk to first degree relatives is increased to six times that of the general population.

Types of hyperlipidaemia

	WHO type	Excess
Autosomal dominant:		
Familial hypercholesterolaemia	IIa, IIb	LDL
Familial combined hyperlipidaemia	IIa, IIb, IV	LDL, VLDL
Familial hypertriglyceridaemia	V, VI	VLDL, CM
Autosomal recessive:		
Apolipoprotein C II deficiency	I, V	CM, VLDL
Polygenic:		
Common hypercholesterolaemia	IIa	LDL

LDL=Low density lipoprotein; VLDL=very low density lipoprotein; CM=chylomicrons.

Lipoprotein abnormalities that increase the risk of heart disease may be secondary to dietary or other factors, but often follow multifactorial inheritance. Familial hypercholesterolaemia (type II hyperlipoproteinaemia) is dominantly inherited and may account for up to 10% of all early coronary heart disease. One in 500 of the general population is estimated to be heterozygous for the mutant gene. The risk of ischaemic heart disease increases with age in heterozygous subjects, who may also have xanthomas. Severe disease, often presenting in childhood, is seen in homozygous subjects.

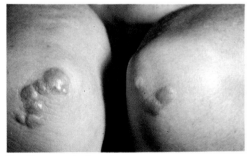

Xanthomas at elbows.

High circulating Lp(a) lipoprotein concentration is also associated with coronary heart disease, and a population attributable risk of 28% for myocardial infarction in men aged under 60 has been suggested. This association may explain much of the familial aggregation of coronary heart disease in the absence of monogenic hyperlipidaemia, although the mechanism underlying the association is not known.

Risk of coronary heart disease from genes for:

- Low density lipoprotein (LDL) receptor
- Lp(a) lipoprotein
- Angiotensin converting enzyme (ACE)

Familial aggregations of early coronary heart disease also occur in people without any detectable abnormality in lipid metabolism. Risks to relatives will be high, and known environmental triggers should be avoided. Molecular genetic studies may lead to more precise identification of subjects at high risk by identifying the major genes responsible. Potential candidate genes include the angiotensin converting enzyme (ACE) gene, and those involved with lipoprotein structure, thrombogenesis, thrombolysis or fibrinolysis, regulation of blood flow through the coronary arteries, regulation of blood pressure, and early development of the coronary arteries.

Schizophrenia and affective psychoses

Overall incidence and empirical risk of recurrence (percentage) in schizophrenia and affective psychosis according to affected relative

	Schizophrenia	Affective psychosis
Incidence in general population	1	3
Sibling	9	13
One parent	13	15
Both parents	40	
Monozygous twin	40	70
Dizygous twin	10	15
Second degree relative	3	5

A strong familial tendency is found in both schizophrenia and affective disorders. The importance of genetic rather than environmental factors has been shown by reports of a high incidence of schizophrenia in children of affected parents and concordance in monozygotic twins, even when they are adopted and reared apart from their natural relatives. The same is true of manic depression. Empirical values for lifetime risk of recurrence are available for counselling, and the burden of the disorders needs to be taken into account. Both polygenic and single major gene models have been proposed to explain genetic susceptibility. A search for linked biochemical or molecular markers in large families with many affected members has so far failed to identify any major susceptibility genes.

Congenital malformations

Risk of recurrence in siblings for some common congenital malformations

	Risk (%)
Anencephaly or spina bifida	5*
Congenital heart disease	1–4
Cleft lip and palate	4
Cleft palate alone	2
Renal agenesis	3
Pyloric stenosis	2–10†
Congenital dislocated hip	1–11†
Club foot	3
Hypospadias	10
Cryptorchidism	10
Tracheo-oesophageal fistula	1
Exomphalos	<1

* Risk reduced by periconceptional supplementation with folic acid.
† Risk affected by sex of index case or sibling, or both.

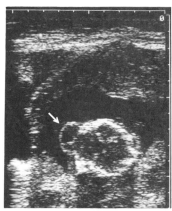

Occipital encephalocele detected by prenatal ultrasonography.

Mendelian, chromosomal, or teratogenic causes can be identified for many syndromes of multiple congenital abnormalities, and improved cytogenetic and DNA techniques are elucidating the cause in others. Some malformations are non-genetic, such as the amputations caused by amniotic bands after early rupture of the amnion. Many isolated congenital malformations, however, follow multifactorial inheritance, and the risk of recurrence depends on the specific malformation, its severity, and the number of affected people in the family. Decisions to have further children will be influenced by the fact that the risk of recurrence is generally low and that surgery for many isolated congenital malformations is successful. Prenatal ultrasonography may identify abnormalities requiring emergency neonatal surgery or severe malformations that have a poor prognosis, but it usually gives reassurance about the normality of a subsequent pregnancy.

Mental retardation

Risk of recurrence for severe non-specific mental retardation according to affected relative

	Risk
One sibling	1 in 35
One sibling with consanguineous parents	1 in 7
Two siblings	1 in 4
One parent	1 in 10
One sibling, one parent	1 in 5
Both parents	1 in 2
Male sibling, maternal uncle or male cousin	X linked

Intelligence is a polygenic trait, and mild mental retardation (intelligence quotient 50–70) represents the lower end of the normal distribution of intelligence. The intelligence quotient of offspring is likely to lie around the mid-parental mean. One or both parents of a mildly retarded child are often retarded themselves and have other retarded children. Intelligent parents with one mildly retarded child are unlikely to have another similarly affected child.

By contrast, the parents of a child with severe mental retardation (intelligence quotient <50) are usually of normal intelligence. A specific cause is more likely when the retardation is severe and may include chromosomal abnormalities and genetic disorders. The risk of recurrence depends on the diagnosis but in severe non-specific retardation is about 3% for siblings, increasing to 25% after the birth of two affected children.

The illustration of encephalocele was reproduced by kind permission of Dr Sylvia Rimmer, St Mary's Hospital, Manchester. The illustration of the twin fetuses discordant for Down's syndrome was reproduced by kind permission of Professor Dian Donnai, St Mary's Hospital, Manchester.

GENETICS OF CANCER

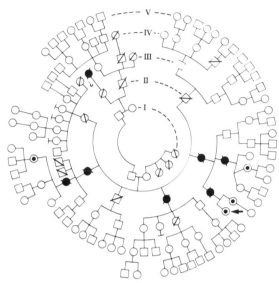

Autosomal dominant inheritance of ovarian adenocarcinoma.

Development of cancer is related to environmental mutagens, somatic mutation, and genetic predisposition. Molecular studies have shown that at least two mutational events are required for the development of malignancy. One of these mutations is inherited in familial cancer. Chromosomal translocations may be markers for (or the cause of) certain neoplasms, and various oncogenes have been implicated.

Though the risk of a common cancer occurring in relatives of an affected person is low, familial aggregations that cannot be explained by environmental factors alone exist in some neoplasms, such as breast, ovarian, and bowel cancers. In occasional families a predisposition to a combination of common cancers is inherited as an incompletely penetrant autosomal dominal trait. Several mendelian syndromes are also associated with a high risk of malignancy, and in many of these the genes responsible have been mapped or cloned.

●	Affected females
⊙	Females at up to 50% risk having undergone prophylactic oophorectomy

Mechanisms of tumorigenesis

The genetic basis of both sporadic and inherited cancers has been confirmed by molecular studies. The three main classes of genes known to predispose to malignancy are oncogenes, tumour suppressor genes, and genes involved in DNA repair. In addition, specific mutagenic defects due to environmental carcinogens and viral infections (notably hepatitis B) are being identified.

Oncogenes are genes that can cause malignant transformation of normal cells. They were first recognised as viral oncogenes (v-onc) carried by RNA viruses. These retroviruses incorporate a DNA copy of their genomic RNA into host DNA and cause neoplasia in animals. Sequences homologous to those of viral oncogenes were subsequently detected in the human genome and called cellular oncogenes (c-onc). More than 60 such proto-oncogenes have been described, and their normal function is in control of cell growth and differentiation. Mutation in a proto-oncogene results in altered, enhanced, or inappropriate expression of the gene product leading to neoplasia. Oncogenes act in a dominant fashion in tumour cells in that mutation in one copy of the gene is sufficient to cause neoplasia. Proto-oncogenes may be activated by point mutations, but also by mutations that do not alter the coding sequence, such as gene amplification or chromosomal translocation. Most proto-oncogene mutations occur at a somatic level, causing sporadic cancers. A notable exception is the germine mutation in the ret oncogene responsible for dominantly inherited multiple endocrine neoplasia type IIa.

Examples of proto-oncogenes implicated in human malignancy

Proto-oncogene	Molecular abnormality	Disorder
myc	Translocation 8q24	Burkitt's lymphoma
abl	Translocation 9q34	Chronic myeloid leukaemia
mos	Translocation 8q22	Acute myeloid leukaemia
myc	Amplification	Carcinoma of breast, lung, cervix, oesophagus
N-myc	Amplification	Neuroblastoma, small cell carcinoma of lung
K-ras	Point mutation	Carcinoma of colon, lung, and pancreas; melanoma
H-ras	Point mutation	Carcinomas of genitourinary tract, thyroid

Tumour suppressor genes normally act to control cell proliferation, and loss or inactivation is associated with tumorigenesis. At the cellular level these genes act in a recessive fashion, as loss of activity of both copies of an autosomal gene is required for malignancy to develop. Tumour suppressor gene inactivation occurs in both sporadic and hereditary cancers connected with this class of gene.

Cloned genes in dominantly inherited cancers

Disorder	Gene symbol	Gene type*	Chromosomal localisation
Basal cell naevus syndrome	BCNS	TS	9q31
Familial adenomatous polyposis	APC	TS	5q21
Familial breast-ovarian cancer	BRAC1	TS	17q21
	BRAC2	TS	13q12-13
Familial melanoma	MLM	TS	9q21
Li-Fraumeni syndrome	P53	TS	17p13
Multiple endocrine neoplasia 2A	MEN 2A	Onc	10q11
Neurofibromatosis type 1	NF1	TS	17q11
Neurofibromatosis type 2	NF2	TS	22q12
Retinoblastoma	RB1	TS	13q14
Tuberous sclerosis	TSC2	TS	16p13
von Hippel-Lindau disease	VHL	TS	3p25
Wilms's tumour	WT1	TS	11p13

* TS=tumour suppressor, Onc=oncogene.

Chromosomal localisation of genes for hereditary cancer that are not cloned in 1996

Disorder	Gene designation	Chromosomal localisation
Beckwith-Wiedemann syndrome	BWS	11p15
Multiple endocrine neoplasia 1	MEN1	11q13
Neuroblastoma	NB	1q36
Renal cell carcinoma	RCC	3p14
Tuberous sclerosis	TSC1	9q34
Tylosis	TY	17q

These two boxes were devised largely from Knudson AG. All in the (cancer) family. *Nature Genetics* 1993; 5: 103–4.

Family with autosomal dominant familial adenomatous polyposis indicating individuals at risk and requiring screening.

Tumorigenesis is a multistage process consisting of two or more initiating events, often involving both oncogenes and tumour suppressor genes. The impact of inheriting a cancer predisposing mutation will be greater in tumours that involve fewer, rather than many, such events. The most commonly altered gene in human cancers is the tumour suppressor gene P53, which occurs in about 70% of all tumours, whereas mutations in the ras oncogene occur in about one third of all cancers. Inherited forms of the common cancers constitute a small proportion of all cases, but are important as they facilitate the localisation and cloning of mutant genes that may also be involved in somatic mutation in sporadic cases. Inherited breast cancer, for example, constitutes only about 5% of all breast cancer cases, but is nevertheless one of the most common of all genetic cancers. The hereditary breast cancer locus (BRCA1) on chromosome 17q is implicated in all hereditary breast and ovarian cancer, about half of hereditary breast cancer alone, and probably has a significant role in the much more common sporadic forms of breast cancer.

The precise mechanism by which oncogenes and tumour suppressor genes act and interact is not known. Interestingly, somatic mutations in the tumour suppressor gene 53 are often found in sporadic carcinoma of the colon, but germline mutation of P53 (responsible for Li-Fraumeni syndrome) seldom predisposes to colonic cancer. Similarly, lung cancers often show somatic mutations of the retinoblastoma (RB1) gene, but this tumour does not occur in individuals who inherit germline RB1 mutations. These genes may play a greater part in progression of tumours than in initiation of cancer. In Lynch cancer family syndrome type 2 (LCFS2) a proportion of cases are linked to a gene on chromosome 2. However, the colonic cancers commonly associated with LCFS2 show somatic mutations similar to those found in sporadic colon cancers, that is in the adenomatous polyposis coli (APC), oncogene K-ras, tumour suppressor P53, and deleted-in-colon-cancer (DCC) genes. This suggests that the LCFS2 gene on chromosome 2 may be acting as a mutagenic rather than a tumour suppressor gene.

There now exists the possibility of gene therapy for cancers, and many of the protocols approved for genetic therapy in the United States are for patients with cancer. Several approaches are being investigated, including virally directed enzyme prodrug therapy, the use of transduced tumour infiltrating lymphocytes that produce toxic gene products, modifying tumour immunogenicity by inserting genes, or the direct manipulation of crucial oncogenes or tumour suppressor genes.

Chromosomal abnormalities in malignancy

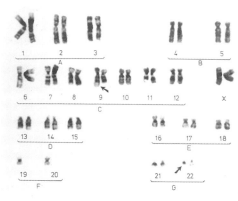

9;22 Translocation in chronic myeloid leukaemia. (One chromosome 19 and one chromosome 20 have also been lost from cell line.)

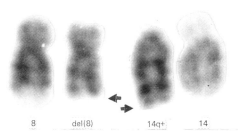

8;14 Translocation in Burkitt's lymphoma.

Structural chromosomal abnormalities are well documented in leukaemias and lymphomas and are used as prognostic indicators. They are also evident in solid tumours—for example, an interstitial deletion of chromosome 3 occurs in small cell carcinoma of the lung. More than 100 chromosomal translocations are associated with carcinogenesis, which in many cases is caused by ectopic expression of chimeric fusion proteins in inappropriate cell types. In addition, chromosome instability is seen in some autosomal recessive disorders that predispose to malignancy, such as ataxia telangiectasia, Fanconi's anaemia, xeroderma pigmentosum, and Bloom syndrome.

Philadelphia chromosome

The Philadelphia chromosome, found in blood and bone marrow cells, is a deleted chromosome 22 in which the long arm has been translocated on to the long arm of chromosome 9 and is designated t(9;22) (q34;q11). The translocation occurs in 90% of patients with chronic myeloid leukaemia, and its absence generally indicates a worse prognosis. The Philadelphia chromosome is also found in 10–15% of acute lymphocytic leukaemias, when its presence indicates a poor prognosis.

Burkitt's lymphoma

Burkitt's lymphoma is common in children in parts of tropical Africa. Infection with Epstein-Barr (EB) virus and chronic antigenic stimulation with malaria both play a part in the pathogenesis of the tumour. Most lymphoma cells carry an 8;14 translocation or occasionally a 2;8 or 8;22 translocation. The break points involve the cellular oncogene c-myc on chromosome 8 at 8q24, the immunoglobulin heavy chain gene on chromosome 14, and the K and λ light chain genes on chromosomes 2 and 22 respectively. Altered activity of the oncogene when translocated into regions of immunoglobulin genes that are normally undergoing considerable recombination and mutation probably plays an important part in the development of the tumour.

Cancer family syndromes

Types of tumour in cancer family syndromes			
Lynch type 2		**Li-Fraumeni**	
Colon	Stomach	Breast	Adrenal
Endometrium	Pancreas	Sarcoma	Embryonal
Ovary	Genitourinary tract	Brain	
Breast		Leukaemia	

Families can be identified in which many relatives develop malignancies, although the tumours may be of different origin. The tumours develop earlier than usual, they may be multifocal, and more than one type of primary tumour may occur in the same person. The two main types identified are the Li-Fraumeni syndrome due to mutations in the P53 suppressor gene and Lynch type 2 which has been shown to be due to mutations in 4 genes involved in mismatch DNA repair. The predisposition to cancer in these rare families behaves as an autosomal dominant trait with 80–90% penetrance so that about 40% of descendants of the proband develop the associated cancers. Identification of the specific mutation in a given family enables predictive genetic testing.

Mendelian cancer syndromes

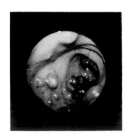

Colonic polyps in familial adenomatous polyposis.

Multiple polyposis syndromes

Familial adenomatous polyposis follows autosomal dominant inheritance and carries a high risk of malignancy necessitating prophylactic colectomy. The presentation may be with adenomatous polyposis as the only feature or as the Gardener phenotype in which there are extracolonic manifestations including osteomas, epidermoid cysts, upper gastrointestinal adenocarcinomas (especially duodenal), and desmoid tumours that are often retroperitoneal. Family members at risk should be screened with regular colonoscopy. Detecting congenital hypertrophy of the retinal pigment epithelium, which occurs in familial adenomatous polyposis, has been used as a method of early identification of gene carriers. The adenomatous polyposis coli gene responsible for familial adenomatous polyposis has been cloned. Linkage analysis and mutation detection in affected families now provides a predictive test to identify gene carriers.

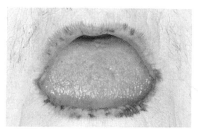

Pigmentation of lips in Peutz-Jehgers syndrome.

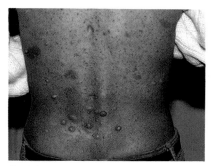

Peripheral neurofibromatosis.

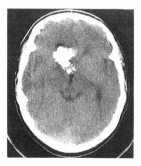

Heavily calcified intracranial hamartoma in tuberous sclerosis.

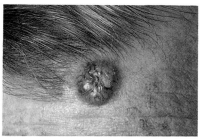

Basal cell carcinoma.

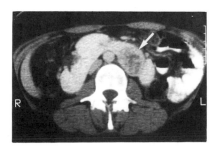

Renal carcinoma in horseshoe kidney on computed tomography in von Hippel-Lindau disease.

In Peutz-Jeghers syndrome hamartomatous gastrointestinal polyps, which may bleed or cause intussusception, are associated with pigmentation of the buccal mucosa and lips. The polyps are usually benign, but malignant degeneration occurs in up to 30–40% of cases. Ovarian, breast, and endometrial tumours also occur in this dominant syndrome.

Neurofibromatosis

Neurofibromatosis type 1 (NF1, peripheral neurofibromatosis, von Recklinghausen's disease) is one of the commonest disorders inherited as an autosomal dominant trait, with an incidence of 1 in 3000. The diagnostic criteria for NF1 include the presence of café au lait patches, cutaneous neurofibromas, and lisch nodules in the iris. Benign optic gliomas and spinal neurofibromas may also occur. Malignant tumours, mainly neurofibrosarcomas or embryonal tumours, occur in 5% of affected people. The gene for NF1 on chromosome 17 has been cloned, but so far mutations have been identified in only a minority of cases.

Neurofibromatosis type 2 (NF2, central neurofibromatosis) is also dominantly inherited. The main feature of NF2 is bilateral acoustic neuromas (vestibular schwannomas). Spinal tumours and intracranial meningiomas occur in over 40% of cases. Surgical removal of VIIIth nerve tumours is difficult and prognosis for this disorder is often poor. The NF2 gene on chromosome 22 has been cloned and various mutations, deletions, and translocations have been identified allowing presymptomatic screening and prenatal diagnosis within affected families.

Tuberous sclerosis

Tuberous sclerosis is an autosomal dominant disorder, very variable in its manifestation, that can cause epilepsy and severe retardation in affected children. Hamartomas of the brain, heart, kidney, retina, and skin may also occur, and their presence indicates the carrier state in otherwise healthy family members. Sarcomatous malignant change is possible but uncommon. Linkage has been shown with chromosome 9 markers in some families, and a chromosome 16 gene mutation is the cause in others.

Naevoid basal cell carcinoma

The cardinal features of the naevoid basal cell carcinoma syndrome, an autosomal dominant disorder delineated by Gorlin, are basal cell carcinomas, jaw cysts, and various skeletal abnormalities, including bifid ribs. Other features are macrocephaly, tall stature, palmar pits, calcification of the falx cerebri, ovarian fibromas, medulloblastomas, and other tumours. The skin tumours are usually bilateral and symmetrical, appearing over the face, neck, trunk, and arms during childhood or adolescence. Malignant change usually occurs, especially after the second decade, and removal of the tumours is therefore indicated. The gene on chromosome 9 has been cloned and is homologous to a drosophila developmental gene called patch.

von Hippel-Lindau disease

In von Hippel-Lindau disease haemangioblastomas develop throughout the brain and spinal cord, characteristically affecting the cerebellum and retina. Renal, hepatic, and pancreatic cysts also occur. The risk of clear cell carcinoma of the kidney is high and increases with age. Phaeochromocytomas occur but are less common. The syndrome follows autosomal dominant inheritance, and yearly screening is recommended for affected family members or those at risk to permit early treatment of problems as they arise. The gene on chromosome 3 has been cloned and identification of mutations allows predictive testing in the majority of families.

Main types of multiple endocrine neoplasia

I	IIa	IIb
Parathyroid	Medullary thyroid	Medullary thyroid
Pancreatic islet cell	Phaeochromocytoma	Phaeochromocytoma
Pituitary	Parathyroid	Mucosal neuromas
Adrenal cortex		
Thyroid		

Childhood tumours

Percentage risk of retinoblastoma in child according to affected relative

	Retinoblastoma
Bilateral tumours:	
Parent	45
Sibling	8
Unilateral tumour:	
Parent and another relative	40
Parent	2
Two siblings	40
One sibling	<1

Two stages of tumour generation

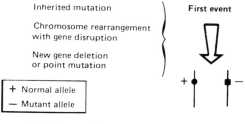

Inherited mutation	First event
Chromosome rearrangement with gene disruption	
New gene deletion or point mutation	

+ Normal allele
– Mutant allele

Loss of normal chromosome and duplication of abnormal chromosome	Second event
Recombination between chromosomes in mitosis	
New gene deletion or point mutation	

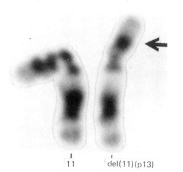
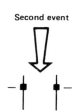

Deletion of chromosome 11 at band 11p13 in patient with Wilms's tumour.

Multiple endocrine neoplasia syndromes

Multiple endocrine neoplasia syndromes

Two main types of the multiple endocrine neoplasia syndrome exist and both follow autosomal dominant inheritance with reduced penetrance. Many affected people have involvement of more than one gland, and first degree relatives in affected families should be periodically screened to detect presymptomatic tumours. Abnormal calcitonin secretion in the type IIa and IIb syndrome, for example, can be detected by calcium or pentagastrin provocation tests, permitting curative thyroidectomy before the tumour cells extend beyond the thyroid capsule. Linkage between the type IIa syndrome and DNA markers on chromosome 10 has led to cloning the gene. Type I multiple endocrine neoplasia has been linked to chromosome 11, but this gene has not yet been isolated.

Retinoblastoma

Most unilateral tumours are sporadic whereas virtually all bilateral tumours are hereditary. Inheritance follows an autosomal dominant pattern with incomplete penetrance; about 80–90% of children inheriting the abnormal gene will develop retinoblastomas. Tumours may occasionally regress spontaneously leaving retinal scars, and parents of an affected child should be examined carefully. Second malignancies occur in up to 15% of survivors in familial cases. In addition to tumours of the head and neck caused by local irradiation treatment, other associated malignancies include sarcomas (particularly of the femur), breast cancers, pinealomas, and bladder carcinomas.

A deletion on chromosome 13 found in a group of affected children, some of whom had additional congenital abnormalities, enabled localisation of the retinoblastoma gene to chromosome 13q14. The esterase D locus is closely linked to the retinoblastoma locus and has been used as a marker to identify gene carriers in affected families, as have other closely linked DNA probes. The retinoblastoma gene has now been cloned. It is a very large gene and specific mutations are not easy to identify.

Molecular studies indicate that two events are involved in the development of the tumour, consistent with Knudson's original "two hit" hypothesis. In bilateral tumours the first mutation is inherited and the second is a somatic event with a likelihood of occurrence of almost 100% in retinal cells. In unilateral tumours both events probably represent new somatic mutations. The retinoblastoma gene is therefore acting recessively as a tumour suppressor gene.

Wilms's tumour

A "two hit" mutation model has also been proposed for Wilms's tumour. Wilms's tumours are usually unilateral, and the vast majority are sporadic. About 1% of Wilms's tumours are hereditary, and of these about 20% are bilateral. Wilms's tumour is associated with aniridia, genitourinary abnormalities, and mental retardation (WAGR syndrome) in a small proportion of cases. Identification of an interstitial deletion of chromosome 11 in such cases localised a susceptibility gene to chromosome 11p13. The Wilms's tumour gene, WT1, at this locus has now been cloned and acts as a tumour suppressor gene, with loss of alleles on both chromosomes being detected in tumour tissue. A second gene at 11p15 has also been implicated in Wilms's tumour. This gene, the insulin-like growth factor-2 (IGF2), is likely to be the cause of Beckwith-Wiedemann syndrome, an overgrowth syndrome predisposing to Wilms's tumour. However, the familial Wilms's tumour gene appears to be linked elsewhere and there is provisional evidence of linkage on chromosome 17.

Illustrations reproduced by kind permission were: the ovarian adenocarcinoma pedigree, Professor Dian Donnai and Dr D Warrell, St Mary's Hospital, Manchester; familial adenomatous polyposis from the *Slide Atlas of Gastroenterology*. Gower Medical Publishing, 1985 and Dr C Williams, St Mark's Hospital, London; the computed tomograms of tuberous sclerosis, Mr P Richardson, Manchester Royal Infirmary and University of Manchester, and renal carcinoma, Professor Judith Adams, University of Manchester Medical School; and the chromosome 11 deletion and 9;22 and 8;14 translocations, Dr Christine Harrison, Christie Hospital, Manchester.

DYSMORPHOLOGY AND TERATOGENESIS

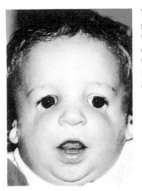

Treacher Collins syndrome: abnormal first branchial arch development giving rise to malar and mandibular hypoplasia with external ear malformations.

Dysmorphology is the study of malformations arising from abnormal embryogenesis. Recognition of patterns of multiple congenital malformations may allow inferences to be made about the timing, mechanism, and aetiology of structural defects. Animal research is providing information about cellular interactions, migration, and differentiation processes and gives insight into the possible mechanisms underlying human malformations. Molecular studies are now identifying defects such as submicroscopic chromosomal deletions and mutations in developmental genes as the underlying cause of some recognised syndromes. Diagnosing multiple congenital abnormalities in children can be difficult but is important to give correct advice about management, prognosis, and risk of recurrence.

Definition of terms

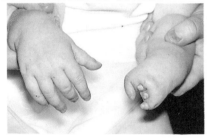

Unilateral terminal transverse defect of the hand occurring as an isolated malformation.

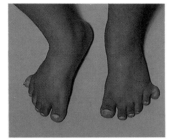

Postaxial polydactyly of the feet in Laurence-Moon-Biedl syndrome (obesity, mental retardation, polydactyly, retinitis pigmentosa, and genital hypoplasia).

Malformation

A malformation is a primary structural defect occurring during development of an organ or tissue. An isolated malformation, such as cleft lip and palate, congenital heart disease, or pyloric stenosis, can occur in an otherwise normal child. Most single malformations are inherited as polygenic traits with a fairly low risk of recurrence, and corrective surgery is often successful. Multiple malformation syndromes comprise defects in two or more systems and are often associated with mental retardation. The risk of recurrence is determined by the aetiology, which may be chromosomal, teratogenic, due to a single gene, or unknown.

Disruption

A disruption defect implies that there is destruction of a part of a fetus that had initially developed normally. Amniotic band disruption after early rupture of the amnion is a well recognised entity, causing constriction bands and amputations of digits and limbs and sometimes more extensive disruptions causing, for example, facial clefts and central nervous system defects. As the fetus is genetically normal and the defects are caused by an extrinsic abnormality the risk of recurrence is small. Interruption of the blood supply to a developing part from other causes will also cause disruption owing to infarction with consequent atresia. The prognosis depends solely on the severity of the physical defect.

Constriction ring with amputation and fusion of digits caused by amniotic bands.

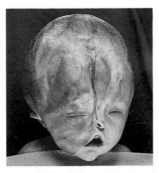

Severe disruption of the face caused by amniotic bands.

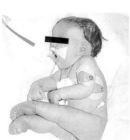

Deformation of legs in newborn infant with hypotonia due to congenital myotonic dystrophy.

Deformation

Deformations are due to abnormal intrauterine moulding and give rise to deformity of structurally normal parts. Deformations usually involve the musculoskeletal system and may occur in fetuses with congenital neuromuscular problems such as spinal muscular atrophy and congenital myotonic dystrophy. Paralysis in spina bifida gives rise to positional deformities of the legs and feet. In these disorders the prognosis is often poor and the risk of recurrence may be high.

Dysmorphology and teratogenesis

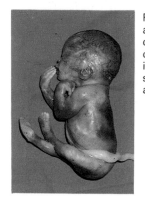

Fetal constraint and deformation due to oligohydramnios in Potter's syndrome (renal agenesis).

Oligohydramnios causes fetal deformation and is well recognised in fetal renal agenesis (Potter's syndrome). The absence of urine production by the fetus results in severe oligohydramnios, which in turn causes fetal deformation and pulmonary hypoplasia. Oligohydramnios caused by chronic leakage of liquor has a similar effect.

A normal fetus may be constrained by uterine abnormalities, breech presentation, or multiple pregnancy. The prognosis is generally excellent, and the risk of recurrence is low except in cases of structural uterine abnormality.

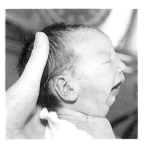

Pierre Robin sequence: mandibular hypoplasia causing cleft palate and respiratory obstruction.

Sequence

The term sequence implies that a series of events occur after a single initiating abnormality, which may be a malformation, a deformation, or a disruption. The features of Potter's syndrome can be classed as a malformation sequence in which the initial abnormality is renal agenesis, which gives rise to secondary deformation and pulmonary hypoplasia. Other examples are the holoprosencephaly sequence and the sirenomelia sequence. In holoprosencephaly the primary developmental defect is in the forebrain, leading to microcephaly, absent olfactory and optic nerves, and midline defects in facial development, including hypotelorism or cyclopia, midline cleft lip, and abnormal development of the nose. In sirenomelia the primary defect affects the caudal axis of the fetus, from which the lower limbs, bladder, genitalia, kidneys, hindgut, and sacrum develop. Abnormalities of all these structures occur in the sirenomelia sequence.

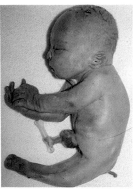

Sirenomelia sequence: fused legs, tail like appendage, absent genitalia, imperforate anus, exomphalos, and renal agenesis.

Associations

Certain malformations occur together more often than expected by chance alone; these are termed associations. The names given to recognised malformation associations are often acronyms of the component abnormalities. Hence the *Vater* association consists of *v*ertebral anomalies, *a*nal atresia, *t*racheo-oesophageal fistula, and *r*adial defects. The acronym *vacterl* has been suggested to encompass the additional *c*ardiac, *r*enal, and *l*imb defects of this association.

Murcs association is the name given to the non-random occurrence of *Mü*llerian duct aplasia, *r*enal aplasia, and *c*ervicothoracic *s*omite dysplasia. In the *Charge* association the related abnormalities include *c*olobomas of the eye, *h*eart defects, *a*tresia choanae, mental *r*etardation, *g*rowth retardation, and *e*ar anomalies.

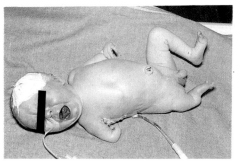

Vater association.

Identification of syndromes

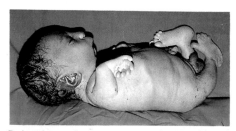

Robert's syndrome: autosomal recessive "pseudothalidomide" syndrome with hypomelia, mid-facial defect, and severe growth deficiency.

Patterns of multiple malformations that occur together constitute syndromes. Recognition of syndromes is important to answer the questions that parents of all babies with congenital malformations ask—namely,

What is it?
Why did it happen?
What does it mean for the child's future?
Will it happen again?

Parents experience feelings of grief and guilt after the birth of an abnormal child, and time spent discussing what is known about the aetiology of the abnormalities may help to alleviate some of their fears. They also need an explanation of what to expect in terms of treatment, anticipated complications, and long term outlook. Accurate

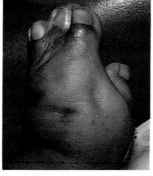

Apert's syndrome: autosomal dominant craniosynostosis with fused digits.

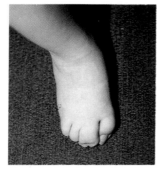

Smith-Lemli-Opitz syndrome: autosomal recessive syndrome with ptosis, anteverted nares, syndactyly of second and third toes, hypospadias, and mental retardation.

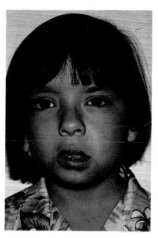
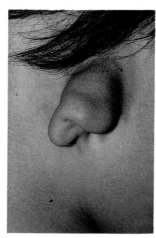

Goldenhar's syndrome (hemifacial microsomia): usually sporadic syndrome with asymmetrical malar, maxillary, and mandibular hypoplasia and microtia.

Stillbirths

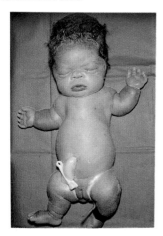
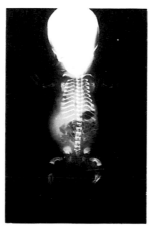

Thanatophoric dwarfism: usually sporadic lethal bone dysplasia.

assessment of the risk of recurrence cannot be made without a diagnosis, and the availability of prenatal diagnosis in subsequent pregnancies will depend on whether there is an associated chromosomal abnormality, a structural defect amenable to detection by ultrasonography, or a biochemical or molecular abnormality.

The assessment of infants and children with malformations requires careful taking of a history and a physical examination. Abnormalities during the pregnancy, including possible exposure to teratogens, should be recorded, as well as the occurrence of any perinatal problems. Parental age and family history may provide clues about the aetiology. Examination of the child should include detailed documentation of the abnormalities present with accurate clinical measurements and photographic records whenever possible, and the investigations required may include chromosomal analysis and molecular, biochemical, or radiological studies.

A chromosomal or mendelian aetiology has been identified for many multiple congenital malformation syndromes. When the aetiology of a recognised multiple malformation syndrome is not known empirical figures for the risk of recurrence derived from family studies can be used, and these are usually fairly low. Consanguineous marriages may give rise to autosomal recessive syndromes unique to a particular family: when more than one child is affected, counselling the couple using the one in four risk of recurrence associated with autosomal recessive inheritance is appropriate.

Numerous malformation syndromes have been identified, and many are extremely rare. Published case reports and specialised texts may have to be reviewed before diagnosis. Computer programs are now available to assist in differential diagnosis, but despite this syndromes in a proportion of children will inevitably remain undiagnosed.

Detailed examination and investigation of malformed stillbirths and fetuses is essential if parents are to be accurately counselled about the cause of the problem, the risk of recurrence, and the availability of prenatal tests in future pregnancies. As with liveborn infants careful documentation of the abnormalities is required with detailed photographic records. Cardiac blood samples and skin biopsy specimens should be taken for chromosome analysis and bacteriological and virological investigations performed. Other investigations, including full skeletal x ray examination and tissue sampling for biochemical studies and DNA extraction, may be necessary. Necropsy will determine the presence of associated internal abnormalities, which may permit diagnosis.

Environmental teratogens

Limb malformation due to intrauterine exposure to thalidomide.

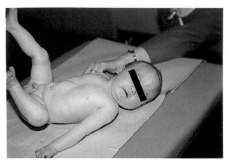

Hypospadias, congenital heart disease, prominent metopic suture (trigonocephaly), and psychomotor retardation in the fetal valproate syndrome.

Examples of teratogens

- *Drugs*
  Alcohol
  Anticonvulsants
    Phenytoin
    Sodium valproate
  Anticoagulants
    Warfarin
  Antibiotics
    Streptomycin
  Treatment for acne
    Tetracycline
    Isotretinoin
  Antimalarials
    Pyrimethamine
  Anticancer drugs
  Androgens

- *Environmental chemicals*
  Organic { mercurials
             solvents

- *Radiation*

- *Maternal disorders*
  Epilepsy
  Diabetes
  Phenylketonuria
  Hyperpyrexia
  Iodine deficiency

- *Intrauterine infection*
  Rubella
  Cytomegalovirus
  Toxoplasmosis
  Herpes simplex
  Varicella-zoster
  Syphilis

Drugs

Identification of drugs that cause fetal malformations is important as they constitute a potentially preventable cause of abnormality. Although fairly few drugs are proved teratogens in humans, and some drugs are known to be safe, the accepted policy is to avoid all drugs if possible during pregnancy. Thalidomide has been the most dramatic teratogen identified, and an estimated 10 000 babies worldwide were damaged by this drug in the early 1960s before its withdrawal.

Alcohol is currently the most common teratogen, and studies suggest that between one in 300 and one in a 1000 infants are affected. Children with the fetal alcohol syndrome exhibit prenatal and postnatal growth deficiency, mental retardation, microcephaly, and characteristic faces with short palpebral fissures, a smooth philtrum, and a thin upper lip. In addition, they have tremulousness owing to withdrawal in the neonatal period.

Treatment of epilepsy during pregnancy presents particular problems as all anticonvulsants are potentially teratogenic. Recognisable syndromes, often associated with mental retardation occur in a proportion of pregnancies exposed to phenytoin and sodium valproate. An increased risk of neural tube defect has been documented with sodium valproate and carbamazepine therapy, and periconceptional supplementation with folic acid is advised. Anticonvulsant therapy during pregnancy may be essential to prevent the risks from grand mal seizures or status epilepticus. Where possible monotherapy using the lowest effective therapeutic dose should be employed. Regardless of treatment, maternal epilepsy itself has been suggested to increase the risk of congenital abnormality in the offspring.

Maternal disorders

Several maternal disorders have been identified in which the risk of fetal malformations is increased including phenylketonuria and diabetes. In phenylketonuria the children of an affected woman will be healthy heterozygotes in relation to the abnormal gene, but if the mother is not returned to a carefully monitored diet before pregnancy the high maternal serum concentration of phenylalanine causes microcephaly in the developing fetus. The risk of congenital malformations in the pregnancies of diabetic women is two to three times higher than that in the general population but may be lowered by good diabetic control before conception and during the early part of pregnancy.

Intrauterine infection

Various intrauterine infections are known to cause congenital malformations in the fetus. Maternal infection early in gestation may cause structural abnormalities of the central nervous system, resulting in neurological abnormalities, visual impairment, and deafness, in addition to other malformations, such as congenital heart disease. When maternal infection occurs in late pregnancy the risk that the infective agent will cross the placenta is higher, and the newborn infant may present with signs of active infection, which include hepatitis, thrombocytopenia, haemolytic anaemia, and pneumonitis.

Rubella embryopathy is well recognised, and the aim of vaccination programmes against rubella virus during childhood is at reducing the number of non-immune girls reaching childbearing age. The presence of rubella specific IgM in fetal or neonatal blood samples identifies babies infected in utero. Cytomegalovirus is a common infection, and 5–6% of pregnant women may become infected. Only 3% of newborn infants, however, have evidence of cytomegalovirus infection, and no more than 5% of these develop subsequent problems. Natural infection with cytomegalovirus does not always confer immunity, and occasionally more than one sibling is affected by intrauterine infection. Unlike with rubella, vaccines against cytomegalovirus or toxoplasma are not available, and although active maternal toxoplasmosis can be treated with drugs such as pyrimethamine, this carries the risk of teratogenesis.

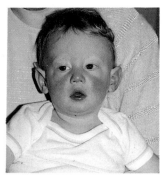

Child with
hepatosplenomegaly, delayed
development, and deafness
due to intrauterine
cytomegalovirus infection.

Herpes simplex infection in the newborn infant is generally acquired at the time of birth, but infection early in pregnancy is probably associated with an increased risk of abortion, late fetal death, prematurity, and structural abnormalities of the central nervous system. Maternal varicella infection may also affect the fetus, causing abnormalities of the central nervous system and cutaneous scars. The risk of a fetus being affected by varicella infection is not known but is probably less than 10%, with a critical period during the third and fourth months of pregnancy. Affected infants seem to have a high perinatal mortality rate.

The illustrations of disruption of the face caused by amniotic bands, congenital myotonic dystrophy, Pierre Robin sequence, sirenomelia sequence, Smith-Lemli-Opitz syndrome, thalidomide malformation, and the valproate syndrome were reproduced by kind permission of Professor Dian Donnai, St Mary's Hospital, Manchester. The illustrations of Potter's syndrome, the Vater association, and thanatophoric dwarfism were reproduced by kind permission of the University of Manchester and Professor Dian Donnai. The illustrations of Treacher Collins syndrome and Goldenhar's syndrome were reproduced from *Dental Update* by permission of Update-Siebert Publications.

PRENATAL DIAGNOSIS

Techniques for prenatal diagnosis

- Ultrasonography
  - —safe
  - —performed mainly in second trimester
- Amniocentesis
  - —procedure risk 0·5–1·0%
  - —performed in second trimester
  - —widely available
- Chorionic villus sampling
  - —procedure risk 2–3%
  - —performed in first trimester
  - —specialised technique
- Cordocentesis
  - —procedure risk 1%
  - —performed in second trimester
  - —specialised technique
- Fetal tissue biopsy
  - —procedure risk <3%
  - —performed in second trimester
  - —very specialised technique
- Embryo biopsy
  - —limited availability and application

Prenatal diagnosis is important in detecting and preventing genetic disease. Two main advances in recent years have been the development of chorionic villus sampling procedures in the first trimester and the application of recombinant DNA techniques to the diagnosis of many mendelian disorders. Various prenatal procedures are available, generally being performed between ten and 20 weeks' gestation. The timing, safety, and accuracy of prenatal tests are important factors that must be considered. Having prenatal tests and waiting for results is stressful for couples. They must be supported during this time and given the results as soon as possible. Many couples who face a high risk of a serious genetic disorder in their children will consider embarking on a pregnancy only if reliable prenatal diagnosis is available. Prenatal testing may also be appropriate for couples in whom the pregnancies are at fairly low risk, often allowing a pregnancy to continue with less anxiety.

Indications for prenatal diagnosis

Prenatal diagnosis occasionally allows prenatal treatment to be instituted but is generally performed to permit termination of pregnancy when a fetal abnormality is detected or to reassure parents when a fetus is unaffected. Pregnancies at risk of fetal abnormality may be identified in various ways. A pregnancy may be at increased risk because of advanced maternal age or abnormal results of biochemical screening; because the couple already have an affected child; or because of a family history of a mendelian disorder or an inherited chromosomal rearrangement. Couples from certain ethnic groups whose pregnancies are at high risk of particular autosomal recessive disorders can be identified before the birth of an affected child by population screening programmes. Screening for carriers of cystic fibrosis is now also possible. In many mendelian disorders, particularly autosomal dominant disorders of late onset and X linked recessive disorders, family studies may be needed to assess the risk to the pregnancy and to determine the feasibility of prenatal diagnosis.

General criteria for prenatal diagnosis

- High genetic risk
- Severe disorder
- Treatment not available
- Reliable prenatal test
- Termination of pregnancy acceptable

Several important factors must be carefully considered before prenatal testing, one of which is the severity of the disorder. For many genetic diseases this is beyond doubt; some disorders lead inevitably to stillbirth or death in infancy or childhood. Perhaps more important, however, are conditions that result in children surviving with severe, multiple, and often progressive, physical and mental handicaps, such as Down's syndrome, neural tube defects, muscular dystrophy, and many of the multiple congenital malformation syndromes.

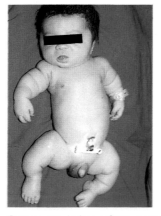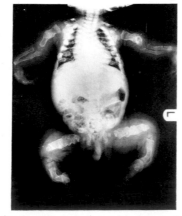

Osteogenesis imperfecta type II (perinatally lethal) can be detected by ultrasonography in second trimester.

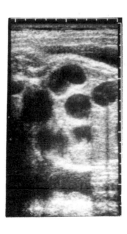

Prenatal detection of jejunal atresia indicating need for neonatal surgery.

The availability of treatment is also important to consider. When treatment is effective termination may not be appropriate and prenatal diagnosis is generally not indicated, unless early diagnosis permits more rapid institution of treatment, reducing illness, complications, and deaths. Phenylketonuria, for example, can be effectively treated after diagnosis in the neonatal period, and prenatal diagnosis, although possible for parents who already have an affected child, may be inappropriate. On the other hand, prenatal diagnosis of congenital malformations amenable to surgical correction is important as it allows the baby to be delivered in a unit with facilities for neonatal surgery and intensive care.

A prenatal test must be sufficiently reliable to permit decisions about a pregnancy. Some conditions can be diagnosed with certainty, others cannot. For example, in mendelian disorders amenable to DNA analysis but for which direct mutation testing is not possible and the biochemical defect in the disorder is not known, the use of linked DNA markers allows a quantified risk to be given for a pregnancy.

As an abnormal result on prenatal testing may lead to termination this course of action must be acceptable to the couples. Careful assessment of their attitudes is important, and even those couples who clearly elect for termination need continued counselling and psychological support afterwards. Couples who do not contemplate termination may still request a prenatal diagnosis to help them to prepare for the outcome of the pregnancy, and these requests should not be dismissed.

Applications of prenatal diagnosis

- Maternal serum screening —α fetoprotein estimation
  —oestriol and human chorionic gonadotrophin estimation
- Ultrasonography —structural abnormalities
- Amniocentesis —α fetoprotein and acetylcholinesterase
  —chromosomal analysis
  —biochemical analysis
- Chorionic villus sampling —DNA analysis
  —chromosomal analysis
  —biochemical analysis
- Fetoscopy —direct examination
  —fetal sampling

Methods of prenatal diagnosis

Some causes of increased maternal serum α fetoprotein concentration

Underestimated gestational age
Threatened abortion
Multiple pregnancy
Fetal abnormality
  Anencephaly
  Open neural tube defect
  Anterior abdominal wall defect
  Turner's syndrome
  Bowel atresias
  Skin defects
Maternal hereditary persistence of α fetoprotein
Placental haemangioma

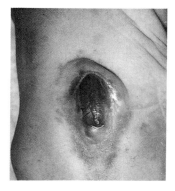

Lumbar meningomyelocele.

Screening of maternal serum

Estimation of maternal serum alfa or fetoprotein concentration in the second trimester has been valuable in screening for neural tube defects—a raised AFP level indicating the need for amniocentesis. Serum screening has now been largely replaced by high resolution ultrasound scanning which detects over 95% of affected fetuses.

In 1992 a combination of maternal serum α fetoprotein, human chorionic gonadotrophin and unconjugated estriol in the second trimester was shown to be an effective screening test for Down's syndrome, providing a composite risk figure taking maternal age into account. Serum screening does not provide a diagnostic test for Down's syndrome, since the results may be normal in affected pregnancies and relatively few women with abnormal serum screening results actually have an affected fetus. However, if 5% of women are selected for diagnostic amniocentesis following serum screening, the detection rate for Down's syndrome is at least 60%, well in excess of the detection rate achieved by offering amniocentesis on the basis of maternal age alone. Serum screening for Down's syndrome is now in widespread use, and modifications enabling first trimester screening are being evaluated.

The isolation of circulating fetal cells, such as nucleated red cells and trophoblasts, from maternal blood offers a potential method for detecting genetic disorders in the fetus by a non-invasive procedure. This method could play an important role in prenatal screening for aneuploidy in the fetus, either as an independent test or in conjunction with other tests such as ultrasonography and biochemical screening.

Prenatal diagnosis

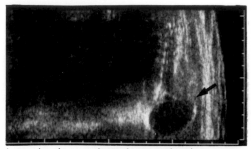

Large lumbosacral meningomyelocele.

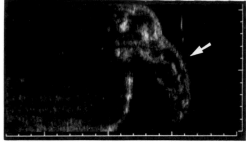

Shortened limb in Saldino-Noonan autosomal recessive bone dysplasia syndrome.

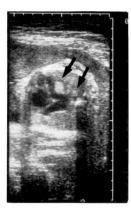

Cardiac leiomyomas in tuberous sclerosis.

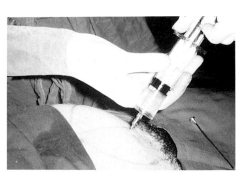

Amniocentesis procedure.

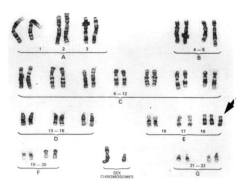

Trisomy 18 karyotype detected by analysis of cultured amniotic cells.

Ultrasonography

Obstetric indications for ultrasonography are well established and include confirmation of viable pregnancy, assessment of gestational age, location of the placenta, assessment of amniotic fluid volume and monitoring fetal growth. Ultrasonography is an integral part of amniocentesis, chorionic villus sampling, and fetal blood sampling and provides evaluation of fetal anatomy during the second and third trimesters.

Disorders such as neural tube defects, severe skeletal dysplasias, and abnormalities of abdominal organs may all be detected by ultrasonography between 17 and 20 weeks' gestation. Centres specialising in high resolution ultrasonography can detect an increasing number of other abnormalities, such as structural abnormalities of the brain, various types of congenital heart disease, clefts of the lip and palate, and microphthalmia. For some fetal malformations the improved resolution of high frequency ultrasound transducers has even enabled detection during the first trimester by transvaginal sonography. Other malformations, such as hydrocephalus, microcephaly and duodenal atresia may not manifest until the third trimester.

Abnormalities may be recognised during routine scanning of apparently normal pregnancies, and this allows the parents to be counselled about the abnormality to consider termination of pregnancy or to make plans for the neonatal management of disorders that are amenable to surgical correction.

Most single congenital abnormalities follow multifactorial inheritance and carry a low risk of recurrence, but the safety of scanning provides an ideal method of screening subsequent pregnancies and usually gives reassurance about the normality of the fetus. Syndromes of multiple congenital abnormalities, however, may follow mendelian patterns of inheritance with high risks of recurrence; for many of these, ultrasonography is the only available method of prenatal diagnosis.

Amniocentesis

Amniocentesis—a well established and widely available method for prenatal diagnosis—is usually performed at 15 to 16 weeks' gestation but can be done a few weeks earlier in some cases. It is reliable and safe, causing an increased risk of miscarriage of around 0·5–1·0%. Amniotic fluid is aspirated directly, with or without local anaesthesia, after location of the placenta by ultrasonography. The fluid is normally clear and yellow and contains amniotic cells, which can be cultured. Contamination of the fluid with blood usually suggests puncture of the placenta and may hamper subsequent analysis. Discoloration of the fluid may suggest impending fetal death.

The main indications for amniocentesis are for chromosomal analysis of cultured amniotic cells in pregnancies at increased risk of Down's syndrome or other chromosomal abnormalities and for estimating α fetoprotein concentration and acetylcholinesterase activity in amniotic fluid in pregnancies at increased risk of neural tube defects, although fewer amniocenteses are now done for neural tube defects because of improved detection by ultrasonography. In specific cases biochemical analysis of amniotic fluid or cultured cells may be required for diagnosing inborn errors of metabolism. Tests on amniotic fluid usually yield results within 7–10 days whereas those requiring cultured cells may take around 2–4 weeks.

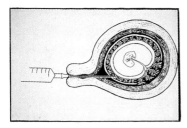

Procedure for transcervical chorionic villus sampling.

Chorionic villus material.

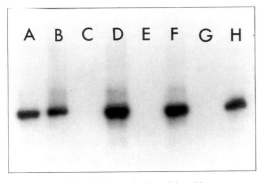

Fetal sexing by DNA analysis with a Y chromosome specific probe.
(Lanes A, B, H, control male; C, G, control female; E, mother, F, father; D, chorionic villus (male).)

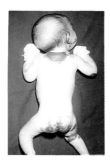

Lethal form of autosomal recessive epidermolysis bullosa can be diagnosed by fetal skin biopsy.

In vitro fertilisation laboratory.

Chorionic villus sampling

Chorionic villus sampling is a technique in which fetally derived chorionic villus material is obtained transcervically with a flexible catheter between ten and 12 weeks' gestation or by transabdominal puncture and aspiration at any time up to term. Both methods are performed under ultrasonographic guidance, and fetal viability is checked before and after the procedure. The risk of miscarriage related to sampling in the first trimester in experienced hands is probably about 2–3% higher than the rate of spontaneous abortions at this time.

Dissection of fetal chorionic villus material from maternal decidua permits analysis of the fetal genotype. The main indications for chorionic villus sampling include the diagnosis of chromosomal disorders and an increasing number of inborn errors of metabolism and conditions amenable to DNA analysis. The advantage of this method of testing is the earlier timing of the procedure, which allows the result to be available by about 12 weeks' gestation, with earlier and easier termination of pregnancy, if required. These advantages are leading to an increased demand for the procedure in preference to amniocentesis, which has important consequences in planning services. If the availability of the procedure is limited, conditions that can be diagnosed only by this method must be given priority, together with high risk cases.

To obtain a prenatal diagnosis in the first trimester it is important to identify high risk situations and counsel couples before pregnancy so that appropriate arrangements can be made and, when necessary, supplementary family studies organised.

Fetal blood and tissue sampling

Fetal blood samples can be obtained directly from the umbilical cord under ultrasound guidance. Blood sampling enables rapid fetal karyotyping in cases presenting late in the second trimester. Indications for fetal blood sampling to diagnose genetic disorders are decreasing with the use of DNA techniques performed on chorionic villus material.

Fetal skin biopsy has proved effective in the prenatal diagnosis of certain skin disorders. Fetal liver biopsy for diagnosis of ornithine transcarbamylace(oic) deficiency has been superseded by DNA analysis and fetoscopy for direct visualisation of the fetus by ultrasonography.

Embryo biopsy

Preimplantation embryo biopsy is now technically feasible. In this method in vitro fertilisation and embryo culture is followed by biopsy of one or two outer embryonal cells at the 6–10 cell stage of development. DNA analysis of a single cell and possibly chromosomal analysis by in situ hybridisation could be performed so that only embryos free of a particular genetic defect would be reimplanted. This method may occasionally be more acceptable than other forms of prenatal diagnosis but has a very limited availability.

Illustrations produced by kind permission were: the ultrasonographic scans of jejunal atresia, meningomyelocele, Saldino-Noonan syndrome, and cardiac leiomyomas, Dr Sylvia Rimmer; meningomyelocele, Professor Dian Donnai; the trisomy 18 karyotype, Dr Lorraine Gaunt; the chorionic villus material, Dr A Andrews; and the autoradiograph of fetal sexing, Mr R Mountford, St Mary's Hospital, Manchester.

TREATMENT OF GENETIC DISORDERS

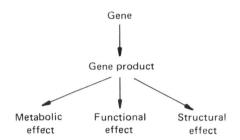

The prevention of inherited disease by means of genetic and reproductive counselling and prenatal diagnosis is often emphasised. Genetic disorders may, however, be amenable to treatment, either symptomatic or potentially curative. Treatment may range from conventional drug or dietary management and surgery to the future possibility of gene therapy. The level at which therapeutic intervention can be applied is influenced by the state of knowledge about the primary genetic defect, its effect, its interaction with environmental factors, and the way in which these may be modified.

Conventional treatment

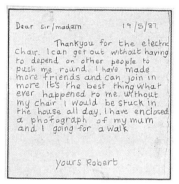

Letter written by boy aged 11 with Duchenne muscular dystrophy.

Increasing knowledge of the molecular and biochemical basis of genetic disorders leads to better prospects for therapeutic intervention and even the possibility of prenatal treatment in some disorders. The primary defect in many disorders, however, is not yet amenable to specific treatment. Conventional treatment aimed at relieving the symptoms and preventing complications remains important and may require a multidisciplinary approach. Management of Duchenne muscular dystrophy, for example, includes neurological and orthopaedic assessment and treatment, physiotherapy, treatment of chest infections and heart failure, mobility aids, home modifications, appropriate schooling, and support for the family, all of which aim at lessening the burden of the disorder. Lay organisations often provide additional support for the patients and their families; the Muscular Dystrophy Group, for example, employs family care officers, who work closely with families and the medical services.

Gene therapy

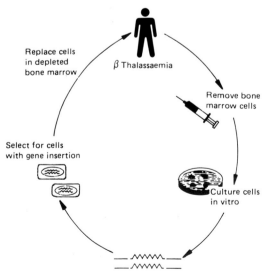

Potential strategy for gene therapy in β thalassaemia.

The prospect of curing genetic disorders with gene therapy is being investigated and has provoked much debate. In theory genetic disease could be cured by manipulating the genome directly to repair a genetic defect or by introducing a functioning donor gene into suitable recipient stem cells. Several experimental strategies have shown that gene transfer is possible. One method is to transfect host cells with retroviruses carrying donor DNA, but this is not yet a therapeutic option in human disease. To be successful the donor DNA needs to be stably incorporated into the host genome and expressed in a controlled manner at adequate levels in the correct cells. Dangers of insertional mutagenesis exist; an inserted donor gene might disrupt a host gene and replace one genetic disorder with another or increase the risk of neoplasia. There are ethical concerns about manipulating the human genome and especially about introducing foreign DNA into germline cells. If a genetic manipulation is not heritable, however, affected descendants will also require treatment.

Gene product replacement

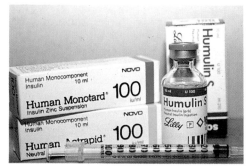

Insulins with human sequence prepared biosynthetically or by enzymatic modification of porcine material.

A simpler alternative to altering the genome is to replace a missing gene product. This strategy is effective when the gene product is a circulatory peptide or protein and is the standard treatment for haemophilia and growth hormone deficiency. The potential for direct replacement of missing intracellular enzymes is being determined experimentally. An alternative method of replacement is that of organ or cellular transplantation, which aims at providing a permanent functioning source of the missing gene product.

When the gene product is needed for metabolism within the central nervous system the blood-brain barrier presents an obstacle to systemic replacement treatment. Another potential problem is the initiation of an immunological reaction to the administered protein by the recipient. Successful production of human gene products, such as insulin and growth hormone, by recombinant DNA techniques may reduce this risk and will ensure adequate supplies for clinical use.

Metabolic manipulation

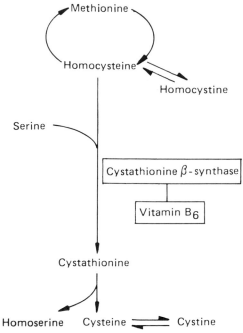

Pathway for homocysteine metabolism: most cases of homocystinuria are due to deficiency of cystathionine β-synthase, which requires vitamin B_6 cofactor.

Many inborn errors of metabolism due to enzyme deficiencies can be treated effectively. Although direct replacement of the missing enzyme is not generally possible, enzyme activity can be enhanced in some disorders. For example, phenobarbitone induces hepatic glucuronyl transferase activity and may lower circulating concentrations of unconjugated bilirubin in the Crigler-Najjar syndrome type 2. Vitamins act as cofactors in certain enzymatic reactions and can be effective if given in doses above the usual physiological requirements; homocystinuria may respond to treatment with vitamin B_6, certain types of methylmalonic aciduria to vitamin B_{12}, and multiple carboxylase deficiency to biotin. It may also be possible to stimulate alternate metabolic pathways; thiamine may permit a switch to pyruvate metabolism by means of pyruvate dehydrogenase in pyruvate carboxylase deficiency.

The clinical features of an inborn error of metabolism may be due to accumulation of a substrate that cannot be metabolised. The classical example is phenylketonuria, in which the absence of phenylalanine hydroxylase results in high concentrations of phenylalanine, causing mental retardation, seizures, and eczema. The treatment consists of limiting dietary intake of phenylalanine to that essential for normal growth. Galactosaemia is similarly treated by a galactose free diet. In other disorders the harmful substrate may have to be removed by alternative means, such as the chelation of copper with penicillamine in Wilson's disease and peritoneal dialysis or haemodialysis in certain disorders of organic acid metabolism. In hyperuricaemia urate excretion may be enhanced by probenecid or its production inhibited by allopurinol, an inhibitor of xanthine oxidase.

Products low in phenylalanine used in dietary management of phenylketonuria.

In another group of inborn errors of metabolism the signs and symptoms are due to deficiency of an end product of a metabolic reaction, and treatment depends on replacing this end product. Defects occurring at different stages in biosynthesis of adrenocortical steroids in the various forms of congenital adrenal hyperplasia are treated by replacing cortisol, alone or together with aldosterone in the salt losing form. Congenital hypothyroidism can similarly be treated with thyroxine replacement. In some disorders, such as oculocutaneous albinism in which a deficiency in melanin production occurs, replacing the end product of the metabolic pathway is, however, not possible.

Treatment of genetic disorders
Environmental modification

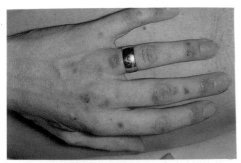

Porphyria cutanea tarda.

The effects of some genetic disorders may be minimised by avoiding known environmental triggers. Certain drugs will precipitate attacks in porphyria, including anticonvulsants, oestrogens, barbiturates, and sulphonamides in acute intermittent porphyria and oestrogens and alcohol in porphyria cutanea tarda. Other drugs, such as primaquine and dapsone, as well as ingesting fava beans cause haemolysis in glucose-6-phosphate dehydrogenase deficiency. Suxamethonium must not be given to people with pseudocholinesterase deficiency, and risks are associated with anaesthesia in myotonic dystrophy.

Exposure to sunlight precipitates skin fragility and blistering in all the porphyrias except the acute intermittent form. Sunlight should also be avoided in xeroderma pigmentosum (a rare defect of DNA repair) and in oculocutaneous albinism because of the increased risk of skin cancer.

Surgical management

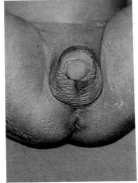

Virilisation of female genitalia in congenital adrenal hyperplasia (21-hydroxylase deficiency).

Surgery plays an important part in various genetic disorders, not only those concerning primary congenital malformations. Virilisation of the external genitalia in girls with congenital adrenal hyperplasia is secondary to excess production of androgenic steroids in utero and requires reconstructive surgery. In some disorders structural complications may occur later, such as the aortic dilatation that may develop in Marfan's syndrome. Surgery may also be needed in genetic disorders that predispose to neoplasia, such as the multiple endocrine neoplasia syndromes, and screening family members at risk permits early intervention and improves prognosis. Occasionally malformations detected during pregnancy, such as posterior urethral valves, may be amenable to prenatal surgical intervention.

The letter was reproduced by kind permission of Robert Little; the illustration of porphyria cutanea tarda by kind permission of Dr T Kingston, Skin Hospital, Salford; and that of congenital adrenal hyperplasia by kind permission of Professor Dian Donnai, St Mary's Hospital, Manchester.

GENE STRUCTURE AND FUNCTION

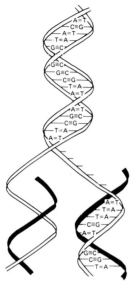

The DNA molecule is fundamental to cell metabolism and cell division as well as providing the basis for inherited characteristics. Nucleic acid, initially called nuclein, was discovered by Friedrich Miescher in 1869, but it was not until 1953 that Watson and Crick produced their model for the double helical structure of DNA and proposed the mechanism for DNA replication. During the 1960s the genetic code was found to reside in the sequence of nucleotides comprising the DNA molecule, a group of three nucleotides coding for an amino acid. The rapid expansion of molecular techniques in the past decade has led to a better understanding of human genetic disease. The structure and function of many genes has been elucidated, and determining the nucleotide sequence of an entire gene is possible. The molecular pathology underlying various disorders is now defined, and DNA analysis can be used for investigating affected families.

DNA structure

Genetic code (RNA)*

First base (5' end)	Second base				Third base (3' end)
	U	C	A	G	
U	Phe	Ser	Tyr	Cys	U
	Phe	Ser	Tyr	Cys	C
	Leu	Ser	Stop	Stop	A
	Leu	Ser	Stop	Trp	G
C	Leu	Pro	His	Arg	U
	Leu	Pro	His	Arg	C
	Leu	Pro	Gln	Arg	A
	Leu	Pro	Gln	Arg	G
A	Ile	Thr	Asn	Ser	U
	Ile	Thr	Asn	Ser	C
	Ile	Thr	Lys	Arg	A
	Met	Thr	Lys	Arg	G
G	Val	Ala	Asp	Gly	U
	Val	Ala	Asp	Gly	C
	Val	Ala	Glu	Gly	A
	Val	Ala	Glu	Gly	G

*Uracil (U) replaces thymine (T) in RNA.

A single strand of DNA consists of a backbone of deoxyribose sugar units linked by phosphate groups. The orientation of the phosphate groups defines the 5' and 3' ends of the molecule. The purine bases adenine (A) and guanine (G) and the pyrimidine bases cytosine (C) and thymine (T) are attached to the deoxyribose units, and their sequence along the molecule constitutes the genetic code. The coding unit, or codon, consists of three nucleotide bases; for example, on the messenger RNA (mRNA) strand, in which uracil (U) replaces thymine, the codon UUC codes for phenylalanine and AGA for arginine; the triplet AUG codes for methionine and also acts as a signal to start protein synthesis; and three triplets (UAA, UAG, and UGA) act as termination signals. As the four bases give 64 possible codon combinations and there are only 20 amino acids most amino acids are specified by at least two codons and the code is said to be degenerate. The DNA code is universal to all organisms with the exception of mitochondrial DNA, which has some different codons.

In the nucleus DNA exists as a double stranded helix in which the order of the bases on one strand is complementary to that on the other. The bases are held together by hydrogen bonds, which allow the strands to separate and rejoin. Adenine is always paired with thymine and cytosine with guanine. This specific pairing is fundamental to DNA replication, during which the two DNA strands separate and each acts as a template for the synthesis of a new strand. The genetic code is therefore maintained during cell division, and as each cell contains an existing and a newly synthesised strand of DNA the process is called semiconservative replication. A damaged DNA strand may be repaired and reconstituted in a similar way.

DNA molecule comprising sugar and phosphate backbone and paired nucleotides joined by hydrogen bonds.

Gene structure and function

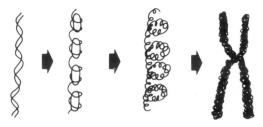

Diagrammatic representation of packaging of DNA into chromosome structure.

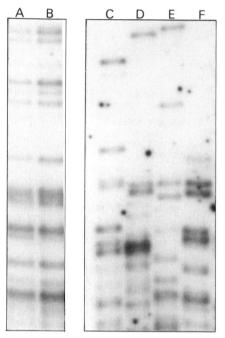

Part of fingerprinting autoradiograph showing bands detected by minisatellite DNA probe in blood samples from monozygous twins (A and B) who have identical bands and four unrelated subjects (C-F) who have different bands.

Transcription and translation

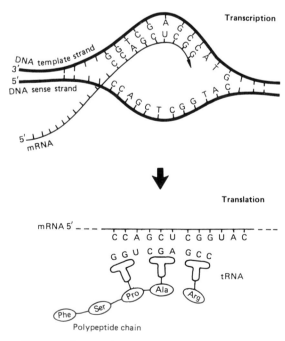

Process of transcription and translation.

Each human somatic cell contains 6×10^9 base pairs of DNA, which is equivalent to about 2m of linear DNA. Packaging of the DNA is achieved by the double helix being coiled around histone proteins to form nucleosomes and then condensed by further coiling into the chromosome structure seen at metaphase. A single cell does not express all of its genes, and active genes are packaged into a more accessible chromatin configuration, which allows them to be transcribed. Some genes are expressed at low levels in all cells and are called housekeeping genes, others are tissue specific and expressed only in certain tissues or cell types.

An estimated 50 000–100 000 pairs of functional genes exist in humans, yet these constitute only a small proportion of total genomic DNA. At least 95% of the genome consists of non-coding DNA, whose function is not clearly defined. Much of this DNA has a unique sequence, but between 30% and 40% consists of repetitive sequences that may be dispersed throughout the genome or arranged as regions of tandem repeats, known as satellite DNA. In tandem repeats the number of times that the core sequence is repeated varies among different people, and this gives rise to hypervariable regions. The repeat motif may consist of a 20–30 base pair sequence in minisatellites or occur as a simple 2 base pair repeat in microsatellites.

The enormous variation occurring in non-coding DNA among different subjects is illustrated particularly well in certain hypervariable minisatellite regions throughout the genome that share a short common core sequence in their variable number of tandem repeats (VNTRs). These hypervariable minisatellite regions are stably inherited and give a pattern of DNA fragments of various sizes on analysis that is unique to each person, forming the basis of the DNA fingerprinting test.

Microsatellite repeats or other DNA variations due to differences in nucleotide sequence that occur close to genes of interest can be used to track genes through families using DNA probes. This approach revolutionised the predictive tests available for mendelian disorders such as Duchenne muscular dystrophy and cystic fibrosis before the genes were isolated and the disease causing mutations identified.

The genetic code carried by DNA is translated into a protein product by means of RNA molecules. The structure of mRNA is similar to that of DNA, except that the sugar backbone is composed of ribose and uracil (U) replaces thymine (T) as one of the bases. One strand of the DNA acts as a template for mRNA synthesis, a process that occurs by pairing of specific bases as it does in DNA synthesis. In experimental systems the reverse of this reaction—the synthesis of complementary DNA (cDNA) using mRNA as a template—is possible with the enzyme reverse transcriptase. This has proved to be an immensely valuable procedure for investigating human genetic disorders as it allows cDNA probes to be produced that correspond exactly to the coding sequence of a human gene.

Messenger RNA is translated into protein in the cytoplasm in association with ribosomes and transfer RNAs (tRNAs). Each tRNA molecule binds a specific amino acid and has three bases forming an anticodon triplet that allows it to bind to a complementary mRNA codon, starting with the initiation signal AUG. Peptide bonds form between the amino acids as the tRNAs are aligned on the ribosome until a stop codon on the mRNA is reached. Several ribosomes associate with each mRNA strand, forming polysomes, thus permitting simultaneous production of several polypeptide chains from one mRNA molecule. The active three dimensional configuration of proteins such as insulin and the collagens is achieved by post-translational modification. Alternative methods of processing and splicing mRNA in certain circumstances permit the production of different protein products from some genes. Gene rearrangement may also occur; rearrangement of the immunoglobulin genes in lymphocytes, for example, is responsible for the enormous structural diversity of antibodies.

Gene structure and function

The coding sequence of a gene is not continuous but is interrupted by varying numbers and lengths of intervening non-coding sequences whose function, if any, is not known. The coding sequences are called exons and the intervening sequences introns. The size and complexity of human genes varies, the largest so far identified is the dystrophin gene, which causes Duchenne muscular dystrophy when it is defective and spans two million base pairs with 14 000 bases of coding sequence distributed among 79 exons.

In addition to the introns, there are non-coding regions of DNA at both 5′ and 3′ ends of genes. There are also regulatory sequences that occur in and around the gene that control its function. In the 5′ flanking region two conserved, or consensus, sequences known as the TATA box and the CAT box, are concerned with controlling mRNA transcription. Methylation of cytosine nucleotides plays a part in regulating gene activity, and enhancer sequences may also be involved, as well as more distant regulatory elements.

Both coding and non-coding DNA sequences in a gene are initially transcribed into mRNA. The sequences corresponding to the introns are then cut out and the exons spliced together to produce mature mRNA. Conserved sequences at the splicing sites enables their recognition in this complex process. Other modifications include the addition of a cap structure at the 5′ end of the mRNA molecule and polyadenylation at the 3′ end.

A genetic mutation may have its effect at any stage in the process of transcription or translation, producing mRNA that is unstable or that cannot be translated into a functional polypeptide. Many different types of mutation are recognised in human genetic disorders, including point mutations, deletions, insertions, rearrangements, and duplications. In different families with the same genetic condition the mutation causing the disorder is not always the same; β thalassaemia, for example, may be due to any of the mutations mentioned above.

Mutations have a wide range of effects. Point mutations, for example, may cause amino acid substitution; disrupt a promotor sequence; alter an initiator codon; generate a terminator codon; or change the codon at a splicing site. Deletions may remove a substantial part of the gene or may alter the reading frame if the number of bases deleted is not a multiple of three, and this has a catastrophic effect on protein synthesis. For example, a deletion in the dystrophin gene that causes a frame shift results in Duchenne muscular dystrophy, whereas one that does not change the reading frame results in the milder Becker's muscular dystrophy.

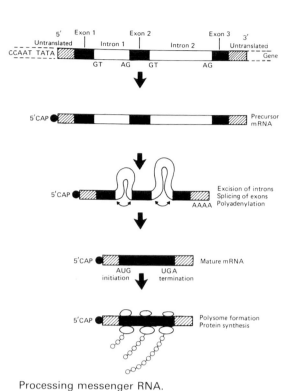

Processing messenger RNA.

The fingerprinting autoradiograph was reproduced by kind permission of Dr A Read, St Mary's Hospital, Manchester.

TECHNIQUES OF DNA ANALYSIS

Molecular genetics laboratory.

DNA analysis is becoming a standard investigation in an increasing number of mendelian disorders. The genetic state of family members and pregnancies at risk can be determined in many conditions, including cystic fibrosis, the haemoglobinopathies, Duchenne muscular dystrophy, and Huntington's disease. The index case is generally diagnosed by means of conventional investigations, such as electrophoresis of haemoglobin in thalassaemia and muscle biopsy in Duchenne muscular dystrophy. DNA studies may clarify the diagnosis in disorders that are associated with specific mutations or gene deletions. The main impact of DNA analysis in clinical practice has been in detecting carriers of X linked recessive disorders, in presymptomatic diagnosis of autosomal dominant disorders, and in prenatal diagnosis of all categories of mendelian disorders. This chapter summarises the standard techniques of DNA analysis that are used in the clinical investigation of affected families. In all health authority regions and health boards throughout the United Kingdom molecular genetics laboratories are associated with departments of clinical genetics and provide DNA analysis as a service to families with particular genetic disorders.

DNA extraction

DNA can be extracted from 10–20 ml whole blood.

Automated DNA extraction.

DNA can be extracted by using standard techniques from any tissue containing nucleated cells, including blood, buccal cells and chorionic villus material. Once extracted, the DNA is stable and can be stored indefinitely so that samples from people with genetic disorders can be collected and saved for the future investigation of other family members. Storing samples has benefited many families whose elderly or affected relatives were no longer living when DNA testing became possible, yet from whom samples were needed to interpret predictive tests.

Restriction enzymes

Preparing restriction enzyme digest reaction.

C C C G G G
G G G C C C
Sma I

C T G C A G
G A C G T C
Pst I

Recognition sequences of two restriction enzymes.

The discovery of bacterial restriction enzymes in 1969 has been important in developing techniques to analyse human DNA. Restriction enzymes recognise specific DNA sequences and cleave double stranded DNA at these sites. Each enzyme has its own recognition sequence and will cut genomic DNA into a series of fragments that can then be analysed. The size of the fragments produced is constant in an individual subject but commonly varies between different subjects because of differences in non-coding DNA sequences. This variation forms the basis of some predictive tests, which are discussed later.

DNA hybridisation

Loading digested DNA samples on to agarose gel.

DNA fragments in gel after electrophoresis stained with ethidium bromide and viewed under nultraviolet light.

Setting up Southern blot: placing filter on to agarose gel.

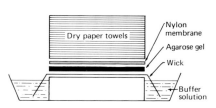

Southern blotting procedure.

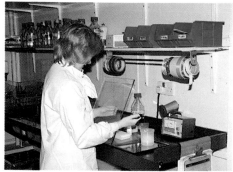

Radiolabelling DNA probe with phosphorus-32.

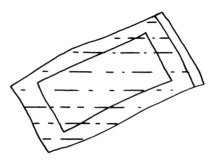

Membrane washed in solution containing radioactive DNA probe during hybridisation reaction.

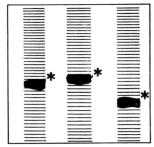

Dense bands (*) on autoradiography indicating hybridisation of probe DNA to homologous sequence in sample DNA.

Electrophoresis

The DNA fragments produced by cutting genomic DNA with restriction enzymes can be ordered according to their size by electrophoresis in an agarose gel. The small fragments migrate faster down the gel than the larger ones, giving a track of DNA fragments of progressively diminishing size. The length of a particular fragment can be determined from the distance of its migration in the gel with reference to marker fragments of known size.

Southern blotting

The DNA fragments in a gel are denatured into single strands and transferred by the technique of Southern blotting on to a nitrocellulose filter or nylon membrane. Fluid rising through the gel transfers the DNA on to the membrane, retaining the alignment of the fragments. Although conventionally performed with a reservoir of buffer solution, the procedure may also be performed without buffer as fluid absorbed directly from the gel by the paper towels is sufficient to permit transfer of the DNA. The DNA binds to the membrane, providing a stable array of DNA fragments that can be analysed by mixing with a DNA probe in a hybridisation reaction. The basis of this reaction is the ability of complementary DNA strands to bind together.

DNA probe hybridisation

A probe is a piece of single stranded DNA, radiolabelled with phosphorus-32, which is used to detect homologous sequences in a sample of genomic DNA. Probes used to study mendelian disorders represent unique sequences that occur only once in the genome and may correspond to the gene of interest, to flanking DNA, or to more distant DNA sequences. Gene specific probes derived from genomic DNA contain both coding and non-coding sequences. Complementary DNA (cDNA) probes are synthesised from the messenger RNA of the gene under study with reverse transcriptase and contain only coding sequences. Oligonucleotide probes, containing around 19 nucleotides, can be synthesised for a specific region of a gene whose sequence is known. Randomly generated probes do not correspond to specific genes but can be used to study families if they are shown to be located close to a gene of interest.

Probes are prepared by cloning the sequence of interest with recombinant DNA techniques. The DNA fragment to be used as a probe is incorporated into vector DNA, usually a bacterial plasmid. The recombinant vector plasmid is then amplified in *Escherichia coli*, allowing large quantities of the probe DNA to be retrieved. The probe is radiolabelled and added to a solution for hybridisation with a membrane blotted with the patient's DNA. The single stranded DNA probe will bind to any DNA fragment on the membrane that has a matching DNA sequence. These fragments can be identified as visible bands in an *x* ray film placed in contact with the membrane (autoradiography).

Techniques of DNA analysis

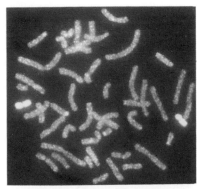

Chromosome spread showing fluorescence in situ hybridisation with probes derived from chromosome 20.

Fluorescence in situ hybridisation

In this technique a DNA probe is labelled with fluorochrome and hybridised direct with a metaphase chromosome spread (see also pages 23 and 27). The presence or absence in an individual of DNA corresponding to the probe DNA can be ascertained by the presence or absence of a fluorescent signal produced by hybridisation to the relevant chromosome, visualised direct using a fluorescence microscope.

DNA amplification

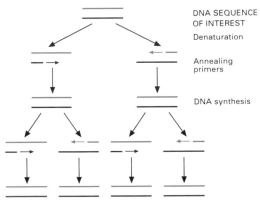

Diagrammatic representation of the polymerase chain reaction (PCR) technique, which uses repeated cycles of denaturation, primer annealing, and DNA synthesis to selectively amplify the DNA sequence of interest.

The polymerase chain reaction (PCR) is used to amplify the DNA of a target sequence many times to allow direct analysis. This enables rapid results to be obtained from very small initial samples, and is particularly useful in prenatal diagnosis. Using the PCR technique it is possible to analyse DNA from single hair bulbs, buccal cells obtained from mouth wash samples, and dried neonatal blood spots. Analysis can also be performed on single cells, making pre-implantation diagnosis of specific genetic disorders a possibility.

The first step in PCR is to heat denature the DNA into two single strands. Two specific oligonucleotide primers that flank the region of interest are then annealed to their complementary strands. In the presence of thermostable polymerase these primers initiate the synthesis of new DNA strands. The cycle of denaturation, annealing, and synthesis repeated 30 times will amplify the DNA from the region of interest 100 000-fold while the quantity of the remaining DNA is unchanged. The amplified sequence can be analysed direct by gel electrophoresis and staining with ethidium bromide.

New techniques

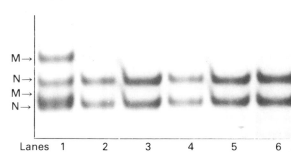

Single strand conformational polymorphism (SSCP) analysis of exon 3 of the cystic fibrosis gene. Lanes 2–6 show migration of the two complementary strands of DNA (N) in the normal gene. Lane 1 shows abnormal migration of DNA strands (M) caused by a point mutation in one copy of the gene, in addition to the normal DNA bands, in a cystic fibrosis heterozygote.

New techniques of DNA analysis are continually being developed which have an appreciable impact on the investigation of genetic disease.

The use of yeast artificial chromosomes (YACS) enables the cloning of large segments of human chromosomes and facilitates the isolation of candidate genes for diseases that have been localised to a particular chromosomal region.

Pulsed field gel electrophoresis (PFGE) permits the separation of large fragments of DNA. This technique is used for long range mapping of the genome and facilitates the identification of gene deletions.

Denaturing gradient gel electrophoresis (DGGE) and single strand conformational polymorphism (SSCP) analysis allow the identification of unknown point mutations and small deletions within genes.

Semiautomated gene sequencing is now a standard procedure in research laboratories, and 200–300 bases can be read at a time. Newer machines linked to computer systems allow fully automated sequence analysis.

RNA analysis is more difficult than DNA analysis and at present has a limited use in service laboratories. Messenger RNA is isolated and a DNA copy made using reverse transcriptase enzyme. The copy DNA which represents only coding sequences, can then be amplified by PCR and sequenced to detect mutations.

The illustration of the fluorescence in situ hybridisation was reproduced by kind permission of Dr Lorraine Grant, Regional Cytogenetics Laboratory, and that of single strand conformational polymorphism analysis by kind permission of Dr A Wallace, Regional Molecular Genetics Laboratory, both at St Mary's Hospital, Manchester.

DNA ANALYSIS IN GENETIC DISORDERS

Examples of mapped autosomal genes

Disorder	Chromosome No
Porphyria cutanea tarda	1
Gaucher's disease	1
von Hippel-Lindau disease	3
Huntington's disease	4
Familial adenomatous polyposis	5
Haemochromatosis	6
21-Hydroxylase deficiency	6
Osteogenesis imperfecta (some forms)	7
Cystic fibrosis	7
Galactosaemia	9
Multiple endocrine neoplasia IIa	10
Sickle cell anaemia and β thalassaemia	11
Acute intermittent porphyria	11
Phenylketonuria (classic)	12
Wilson's disease	13
Retinoblastoma	13
α_1-Antitrypsin deficiency	14
Tay-Sachs disease	15
α Thalassaemia	16
Adult polycystic kidney disease	16
Neurofibromatosis 1 (peripheral)	17
Osteogenesis imperfecta (some forms)	17
Familial hypercholesterolaemia	19
Myotonic dystrophy	19
Alzheimer's disease (familial)	14, 21
Homocystinuria	21
Hurler's syndrome (mucopolysaccharidosis I)	22
Neurofibromatosis 2 (central)	22

The ability to analyse DNA has had an important impact on our understanding of genetic disorders. Many genes have now been cloned and sequenced and the mutations that cause disease identified. Other genetic disorders, for which the genes are not yet isolated, have been mapped to particular chromosomal locations, and this permits predictive testing with linked DNA markers. It has been estimated that the entire human genome will be sequenced and all the important genes mapped by the end of the century as part of the human genome project. This is not unrealistic and promises to be of enormous benefit to families with genetic disorders and to potential gene carriers.

International meetings on human gene mapping, inaugurated in 1973, have been held every two years to document progress. At the first meeting the total number of autosomal genes whose chromosomal location had been identified was 64. The corresponding number of mapped genes had risen to 928 by the ninth meeting in 1987. This tremendous increase reflects the addition of various molecular biological approaches to those of more traditional somatic cell genetics. The total number of mapped X linked loci also rose, from 155 in 1973 to 308 in 1987. It is now estimated that over 2400 genetic disorders have been studied by DNA linkage methods in addition to those localised using other methods.

In this chapter some of the ways in which molecular techniques can be applied clinically are illustrated. In most genetic conditions amenable to prenatal diagnosis by DNA analysis studies must be performed on the family first to identify the particular mutation involved or, if that is not feasible, to discover whether there is a DNA variation that gives an informative pattern. Certain key relatives must be available for testing to make prediction possible. Because chorionic villus sampling is performed at ten weeks of gestation counselling and investigating a family before pregnancy is important to ensure that a couple has time to make fully informed decisions.

Gene localisation

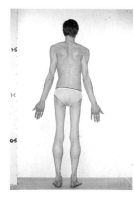

Becker's muscular dystrophy: proximal muscle wasting, winging of scapulae, and pseudohypertrophy of calf and deltoid muscles. The gene for this disorder was localised by DNA linkage studies.

Some disease genes have been localised by finding an affected individual with a chromosomal rearrangement that disrupts the gene. In these cases analysis of the chromosomal break points allows the gene to be cloned and sequenced, as occurred with neurofibromatosis type 1 (NF1) and familial bowel cancer.

More often, however, the first step in locating genes is to perform linkage studies in affected families using polymorphic DNA markers. Using this strategy it is possible to locate any disease gene of interest provided that sufficient numbers of large families are available for study. A DNA marker of known chromosomal location that segregates with the disease in families will identify the location of the gene. Because recombination occurs between homologous chromosomes at meiosis, a DNA marker that is not close to a gene on a particular chromosome or is on a different chromosome will be inherited independently of the gene. The closer the marker is to a gene, the less likely it is that recombination will occur. In practice, markers that show less than 5% recombination with a disease gene are useful in detecting carriers and in prenatal diagnosis. As 1% recombination occurs between loci that are separated by around one million base pairs, the markers may be up to five million bases away from the gene being studied.

Gene tracking

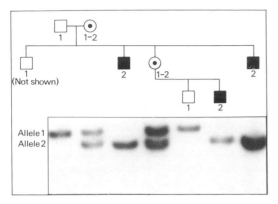

Autoradiograph showing restriction fragment length polymorphism (alleles 1 and 2) detected by X chromosomal DNA probe linked to gene for Becker's muscular dystrophy: disease gene segregates with maternal allele 2, providing a marker for the disorder, subject to recombination.

There are several different types of variation in DNA that can be used as markers to track disease genes through families once they have been shown by linkage studies to occur in or near the gene of interest. Sequence variations (polymorphisms) in non-coding DNA are extremely frequent throughout the genome. When they affect restriction enzyme cleavage sites, DNA fragments of different sizes will be obtained after restriction endonuclease digestion of the DNA. The different alleles produced are called restriction fragment length polymorphisms (RFLPs). A better source of DNA markers is provided by the thousands of clustered repeat sequences known as variable number tandem repeats (VNTRs) that are scattered throughout the genome. Dinucleotide CA repeats (microsatellites) are particularly useful as they are highly polymorphic and identify multiple allele systems, in contrast to the RFLPs which usually identify only two or three variant alleles.

Before gene tracking can be used to provide a predictive test, other family members known to be affected or unaffected must be tested to find an informative DNA marker within the family and to identify which allele is associated with the disease gene in that particular kindred. Though this method of predictive testing is of great value for disorders in which direct mutation analysis is not possible, there is nearly always a margin of error in the results because of a residual risk of recombination having occurred between the marker and gene locus.

Mutation analysis

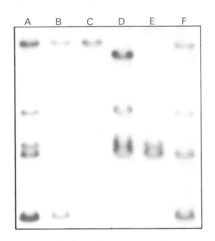

Autoradiographic bands corresponding to exons of dystrophin gene detected by probe Cf56 in DNA digested with *Pst I* restriction endonuclease from boys with Duchenne muscular dystrophy. Lane A: five exons with no deletion; lanes B–F: various exons deleted; lane D: also shows alteration in size of largest exon.

Deletions

Gene deletions are the causal mutations in several disorders including α thalassaemia and some cases of β thalassaemia, haemophilia A, and Duchenne muscular dystrophy. Large deletions can be detected direct by the absence of specific DNA bands on autoradiography after hybridisation with gene specific probes or by the absence of polymerase chain reaction (PCR) products using primers located within the deleted region. With these methods, deletions are most easily identified in X linked disorders, as males have only one X chromosome. In autosomal disorders the deletion is harder to detect because of the presence of a normal copy of the gene on the other chromosome of the autosomal pair. Dosage studies may be used or alternatively large deletions may be detected by fluorescence in situ hybridisation (FISH) techniques, for example in Duchenne muscular dystrophy carrier testing (see pages 23, 52, 56, and 64).

Small deletions can be detected by the PCR technique using primers that amplify across the deletion. In this case two DNA fragments will be amplified, the larger one representing the normal gene sequence, the smaller one representing the deleted gene.

Point mutations

Most disease-causing mutations are simple base substitutions. These point mutations are amenable to direct detection if the mutation is known and affects a recognition site for a restriction enzyme or if specific oligonucleotide probes are available. In sickle cell disease a point mutation changes the codon GAG to GTG in the β globin gene and results in the substitution of valine for glutamic acid in haemoglobin. The mutation alters the recognition site of the restriction enzyme *Mst* II, changing the size of the DNA fragment detected by the β globin gene on autoradiography.

An alternative method for testing for sickle cell disease or its carrier state is to use oligonucleotide probes corresponding to the normal and mutant β globin gene sequences. Each oligonucleotide probe will bind only to its specific genomic counterpart, so that the haemoglobin A probe gives an autoradiographic band only with the normal β globin gene, and the haemoglobin S probe gives a band only with the mutant β globin gene. Both probes hybridise with the DNA from heterozygous subjects.

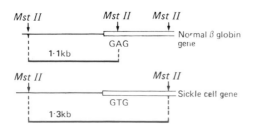

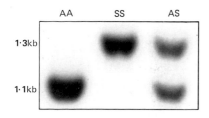

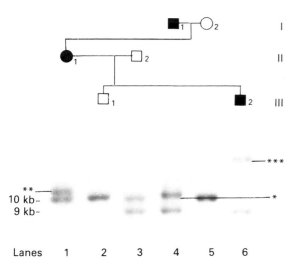

Autoradiograph showing detection of sickle cell mutation by altered size of DNA fragments produced by cleavage with *Mst II* restriction enzyme.

Autoradiograph showing DNA fragments containing the trinucleotide repeat region associated with the myotonic dystrophy protein kinase gene. The 9 kb and 10 kb fragments represent normal alleles. Expansions of variable size are shown in subjects with myotonic dystrophy (lanes 1, 4, and 6). Minimal expansion* (more accurately detected by polymerase chain reaction (PCR) sizing) is present in the minimally affected grandfather (I_2). Further expansion ** is present in the mildly affected mother (II_1). The largest expansion *** is present in the congenitally affected child (III_2).

Point mutations can also be detected using a modified PCR reaction, and this method is now widely used, notably in screening for cystic fibrosis mutations. PCR primers specific to the normal and mutant alleles are generated and used in separate amplification reactions. The presence of normal and mutant alleles is indicated by DNA amplification in the corresponding reaction.

Trinucleotide repeat expansions

Expanding trinucleotide repeats represent a new unstable mutation recently identified. This type of mutation is the cause of three major genetic disorders—fragile X syndrome, myotonic dystrophy, and Huntington's disease—and has also been found in X linked spinobulbar neuronopathy (Kennedy disease) and dominantly inherited spinocerebellar ataxia type 1. In the normal copies of these genes the number of repeats of the trinucleotide sequence is variable. In affected individuals the number of repeats expands outside the normal range. In Huntington's disease the expansion is small, involving a doubling of the number of repeats from 20–35 in the normal population to 40–80 in affected individuals.

In fragile X and myotonic dystrophy the expansion may be very large, and the size of the expansion is often unstable when transmitted from affected parent to child. Severity of the disorder in fragile X and myotonic dystrophy correlates with the size of the expansion: larger expansions causing more severe disease. In myotonic dystrophy the severe congenital form of the disease occurs in a proportion of cases, but only when the mutation is transmitted by an affected mother. This is associated with the presence of the largest trinucleotide expansions in the infant. In fragile X syndrome, large expansions are associated with DNA methylation and have their effect by inactivating the FMR1 (fragile X mental retardation type 1) gene.

Small trinucleotide expansions can be amplified and sized using PCR techniques. It is not possible to amplify across the trinucleotide repeat region in very large expansions, but these can be detected by Southern blot analysis and hybridisdation with a probe located close to the region of the expansion.

The illustration of Becker's muscular dystrophy was reproduced by kind permission of Dr K Cumming, Withington Hospital, Manchester, and the autoradiographs of Duchenne muscular dystrophy deletions by Mr R Mountford and of myotonic dystrophy by Mr S Ramsden, both of St Mary's Hospital, Manchester.

MOLECULAR GENETICS OF SOME COMMON MENDELIAN DISORDERS

Haemoglobinopathies

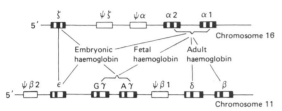

Globin gene clusters on chromosomes 11 and 16. (ψ denotes pseudogenes.)

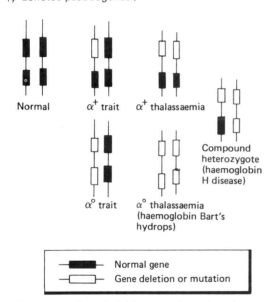

Representation of globin genes in various forms of α thalassaemia.

Cystic fibrosis

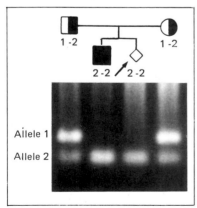

Linkage analysis in prenatal diagnosis of cystic fibrosis using polymerase chain reaction (PCR). The affected child (lane 2) is homozygous for allele 2, which identifies the parental alleles linked to the mutant cystic fibrosis gene. The fetus (lane 3) has inherited identical alleles to the affected child and is therefore predicted to be affected.

The haemoglobinopathies constitute the most common autosomal disorders world wide and have profound effects on the provision of health care in some developing countries. They were among the first disorders to be analysed at a molecular DNA level, partly because the structure of haemoglobin was already well defined and also because fairly pure RNA could be extracted from reticulocytes and used to produce complementary DNA probes. The globin gene clusters on chromosome 16 include two α globin genes and on chromosome 11 a β globin gene.

Various mutations in the β globin gene cause structural alterations in haemoglobin, the most important being the point mutation that produces haemoglobin S and causes sickle cell anaemia. Direct detection of the point mutation (described in the previous chapter) permits early prenatal diagnosis by analysis of DNA extracted from chorionic villus material.

The thalassaemias are due to a reduced rate of production of one or more globin chains, leading to an imbalance in their production. In α thalassaemia production of α globin chains may be absent (α^0) or reduced (α^+). In the α^0 thalassaemia trait both α globin genes are deleted from one chromosome and in the homozygous state all four genes are deleted. In the α^+ thalassaemia trait only one α globin gene is inactivated, either by deletion or mutation, and the other is intact. Deleted genes can be detected direct in the homozygous state by failure of hybridisation with α globin gene probes.

In β thalassaemia over 150 different mutations causing the disorder have been identified, which result in β^0 and β^+ types depending on whether the production of β globin chains is absent or reduced. Major gene deletions are unusual in β thalassaemia, and most mutations entail point mutations or small deletions or insertions. To offer prenatal diagnosis by DNA analysis each individual family must be studied to determine the nature of the mutation. Particular mutations are common in certain populations, and specific oligonucleotide probes are available for prenatal diagnosis in many cases.

Cystic fibrosis is the commonest autosomal recessive disorder in northern Europeans, and about 1 in 20 of the population is a carrier, resulting in a disease incidence of 1 in 2000 live born infants. Cystic fibrosis is a multisystem disorder affecting exocrine gland function, which results in chronic lung disease and malabsorption. The disorder remains incurable, although survival is improved with supportive treatment and advances in molecular medicine offer the prospect of future gene therapy.

Before gene localisation, carrier detection was not possible and prenatal diagnosis could only be offered to pregnancies at high risk—that is, to couples who already had an affected child. This was performed by measuring the activity of microvillar enzymes in amniotic fluid but was associated with false positive and false negative results. Family studies localised the gene to chromosome 7q31 in 1985, and several closely linked markers were soon available for tracking the gene through families. Provided that a family was informative for a polymorphic marker and that a DNA sample from the affected child could be analysed, more accurate prenatal diagnosis could be offered, and carrier testing for affected families became possible (see figure).

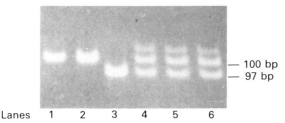

Lanes   1   2   3   4   5   6

— 100 bp
— 97 bp

Polymerase chain reaction (PCR) analysis of △F508 mutation in cystic fibrosis. Lower band represents a 97 base pair PCR product from the mutant allele containing the 3 base pair △F508 deletion. The middle band represents a 100 base pair PCR product from the normal allele. Both bands are present in heterozygotes, together with an upper band that is generated when both normal and mutant DNA strands are present. Lanes 1 and 2: homozygous normal. Lane 3: homozygous affected. Lanes 4, 5, and 6: heterozygous carrier.

Prevalence of the four most common cystic fibrosis mutations*

Mutation	Prevalence (%)*
△F508	79·2
G551D	3·0
G542X	1·4
621 + 1G → T	0·8

* Prevalence related to samples tested in the Regional Molecular Genetics Laboratory, St Mary's Hospital, Manchester.

Huntington's disease

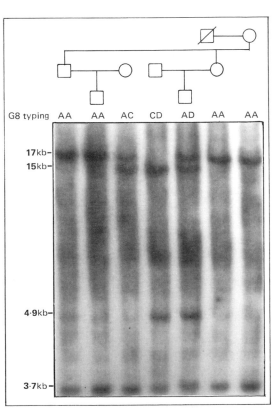

G8 typing   AA   AA   AC   CD   AD   AA   AA

17kb-
15kb-

4·9kb-

3·7kb-

Inheritance of G8 haplotypes in DNA digested with *Hind* III restriction endonuclease (A = 17, 3·7 kilobases (kb); B = 17, 4·9 (not shown); C = 15, 3·7 kb; and D = 15, 4·9 kb).

Molecular genetics of some common mendelian disorders

Most chromosomes carrying the cystic fibrosis mutation in northern Europeans were found to carry a particular combination of marker alleles that were uncommon in normal chromosomes. This linkage disequilibrium between certain marker alleles and the cystic fibrosis gene indicated that most chromosomes carrying the cystic fibrosis gene originated from a single ancestral mutant. This allowed calculation of carrier risk for a person without a family history of cystic fibrosis, but was of limited practical value.

The cystic fibrosis transmembrane conductance regular gene (CFTR) was cloned in 1989. It spans 250 kilobases of DNA and contains 27 exons. The mRNA transcript is 6·5 kilobases in size and encodes a polypeptide of 1480 amino acids that functions as a chloride channel. The first mutation detected involves a 3 base pair deletion in exon 10 of the gene resulting in the loss of a phenylalanine residue at position 508. This mutation is called ΔF508 and accounts for over 70% of all cystic fibrosis mutations in northern Europeans. Over 400 different mutations have been identified. Their frequency varies between populations, and many are very rare. Mutation detection permits definitive carrier testing and prenatal diagnosis in affected families.

Screening for the four most common cystic fibrosis mutations in northern Europeans using polymerase chain reaction (PCR) techniques identifies 85% of all carriers in the general population. The aim of population carrier screening is to identify carrier couples before the birth of their first affected child, and about 72% of such couples can be identified. Using this protocol several carrier screening programmes have been established, both in a primary care setting and in antenatal clinics. A woman found to be a cystic fibrosis carrier can be reassured about the low risk (less than 1 in 500) of having a child with cystic fibrosis if her partner does not carry an identifiable mutation. When both partners are found to be carriers they can be given information about the disease, the 1 in 4 risk to each pregnancy, the availability of precise prenatal diagnosis and offered selective termination of affected pregnancies if desired.

Huntington's disease is an autosomal dominant condition and is one of the most devastating genetic disorders, with onset of involuntary movements, dementia, and personality disorder being variable but commonly occurring between the ages of 35 and 55. Life expectancy averages 15 years from onset of symptoms. Because of the late onset of the disorder many people at risk have had to make reproductive and other important life decisions before knowing whether they will develop the disease themselves or not. The gene for Huntington's disease was linked to a probe called G8 in family studies in 1983, localising the gene to the short arm of chromosome 4. The G8 probe detected polymorphisms with restriction enzyme *Hind* III, and the four haplotypes, designated A, B, C, and D, could be tracked through affected families. Together with other linked probes subsequently identified, this made predictive testing possible for a person at risk, provided that the family structure was suitable and an informative marker could be found. Many people at risk, however, did not want predictive testing in the absence of any effective treatment. The potential for predictive testing did however raise many important ethical issues. The need for careful counselling before and after such tests was recognised, and an internationally agreed code of practice for predictive testing has been devised.

A major concern of many people at risk of developing the disease is that they may transmit the disorder to their children. Linkage analysis provided a "fetal exclusion test" that indicated the risk to a fetus without predicting the genetic state of the parent. The principle of the test is that it showed whether the fetus had inherited a chromosome from the affected or unaffected grandparent through the parent at risk. A chromosome from the affected grandparent confers a 50% risk (the same as the risk to the parent). A chromosome inherited from the unaffected grandparent reduces the risk to that associated with the chance of recombination having occurred between the gene for the

Molecular genetics of some common mendelian disorders

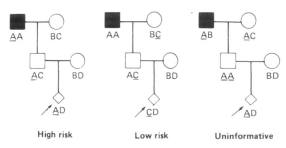

Examples of fetal exclusion test results in Huntington's disease.

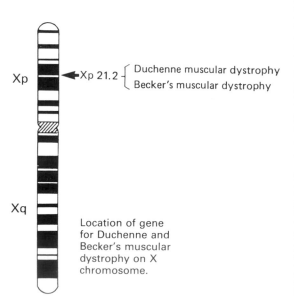

Lanes   1   2   3   4   5   6   7   8

Size range of alleles with trinucleotide expansion associated with Huntington's disease

Size range of normal alleles

Polymerase chain reaction (PCR) analysis of trinucleotide repeat region associated with Huntington's disease. Lanes 2 and 6: size markers. Lanes 1 and 5: normal individuals. Lanes 3, 4, 7, and 8: affected heterozygotes with one allele containing a pathological expansion of the trinucleotide repeat.

disease and the probe being used. One potential problem with this type of test is that if a pregnancy identified as being at 50% risk continues to term and the parent subsequently develops Huntington's disease this indicates that the child has probably also inherited the gene and will develop the disorder.

The genetic mutation responsible for Huntington's disease has now been identified as an expansion of a CAG trinucleotide repeat. Normal genes have between 9 and 34 repeats and 97·5% have fewer than 28 repeats. People with Huntington's disease have expansions of 37–80 repeats. As with other disorders associated with trinucleotide repeats, the expansion is unstable and varies from one generation to the next.

The instability in the Huntington's disease gene is much less than that observed in myotonic dystrophy and fragile X syndrome. The size of the Huntington's disease expansion correlates inversely with age at onset. Repeats of over 55 trinucleotides occur in individuals with age at onset under 30 years, although some early onset cases have shown 35 and 50 repeats. Fewer than 50 repeats have no relation to age of onset, thus repeat number has little predictive value in an individual situation. Greater instability of the expansion is observed with paternal transmission and may explain the association between paternal Huntington's disease and juvenile onset in offspring.

Direct detection of the Huntington's disease mutation has obvious clinical impact, but needs to be used with great care, especially in interpreting the importance of expansions in the intermediate range. DNA testing may now confirm the diagnosis in cases of clinical uncertainty, and definitive predictive tests can potentially be offered to all individuals at risk without the need to perform the family studies previously used in linkage. It remains crucial, however, to consider carefully all the implications of predictive testing before performing DNA analysis.

Duchenne muscular dystrophy

Xp

Xp 21.2 { Duchenne muscular dystrophy
             Becker's muscular dystrophy

Xq

Location of gene for Duchenne and Becker's muscular dystrophy on X chromosome.

Duchenne muscular dystrophy was first described in 1861 and the X linked pattern of inheritance reported in 1943. A milder form of X linked muscular dystrophy identified by Becker in 1955 is now known to be due to a defect in the same gene. Duchenne muscular dystrophy has been reported in girls who have one X chromosome disrupted by a translocation between it and an autosome, with the normal X chromosome being preferentially inactivated. The site of the breakpoint in the cases of the chromosomal translocation is always located in the Xp21 band, which suggested that this was the location of the Duchenne gene. DNA probes identified from this region were shown to be linked to Duchenne muscular dystrophy in family studies in 1983, confirming this localisation. Other probes showing closer linkage have subsequently been identified and used in detecting carriers. Strategies were then devised to obtain DNA probes from within the gene by using DNA from a patient with a chromosomal deletion and from another with a chromosomal translocation. The gene that causes Duchenne and Becker's muscular dystrophy when it is defective has since been cloned and the gene product, dystrophin, identified. The gene spans 2·4 million base pairs (Mb) of DNA and contains 79 exons. The 14 kilobase mRNA transcript encodes dystrophin, a cytoskeletal membrane protein that is expressed in skeletal, cardiac, and smooth muscle as well as in the brain.

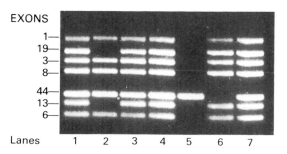

EXONS

Lanes 1 2 3 4 5 6 7

Polymerase chain reaction (PCR) multiplex analysis in Duchenne muscular dystrophy identifying exons 1, 19, 3, 8, 44, 13, and 6. Deletions are present in three samples. Lane 2: patient with deletion including exons 13-19. Lane 5: patient with deletion including exons 1-19. Lane 6: patient with deletion of exon 44.

Polyclonal antibodies raised to the dystrophin protein can be used for histochemical and protein analysis in muscle biopsy samples. Patients with Duchenne muscular dystrophy have little or no demonstrable protein, while patients with Becker's muscular dystrophy have reduced amounts of dystrophin or produce dystrophin of abnormal size. In most cases caused by gene deletion the severity of the disease can be explained by the "frameshift hypothesis." This proposes that a deletion of bases that is not a multiple of three will alter the codon reading frame and produce a non-functional protein. A deletion that does not alter the reading frame results in production of a modified protein with residual function, which causes the milder Becker phenotype.

About 60–70% of males with Duchenne and Becker's muscular dystrophy have deletions of coding sequences in the dystrophin gene as shown by hybridisation with complementary DNA probes. A PCR multiplex reaction involving the analysis of 18 exons can detect 99% of all deletions. Demonstrating a deletion confirms the clinical diagnosis, which can be difficult in mild sporadic cases, and also provides a definitive prenatal diagnostic test for families in which an affected boy has a gene deletion. Chorionic villus sampling in pregnancies at risk allows fetal sexing to be performed by DNA analysis and in male fetuses the presence or absence of a deletion can be investigated. This allows unaffected male pregnancies to continue to term, which was not possible when only fetal sexing was available. Different deletions are seen in different families, and the particular deletion involved needs to be identified to allow appropriate prenatal testing.

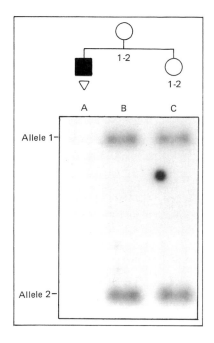

Restriction fragment length polymorphism and deletion detected with probe p20 in DNA digested with *Msp* I restriction endonuclease. (See text for interpretation.) ▽ = Deletion.

DNA hybridisation using an appropriate probe does not necessarily identify female carriers of a gene deletion, as the DNA bands corresponding to the normal gene on one X chromosome mask the presence of a deleted gene on the other. Dosage studies (assessing whether one or two copies of the relevant exons are present) are not reliable enough to permit routine prediction. If a probe that detects a deletion in a family also, however, identifies an associated polymorphism this can be used in assessing carrier state as shown by the autoradiograph opposite. This example illustrates the use of a probe identifying an RFLP, although CA microsatellite markers are now used in preference because their greater variability makes them more informative and they can be studied by PCR. DNA from the affected boy (lane A) shows no bands with probe p20, indicating a gene deletion. His mother (lane B) is heterozygous for the associated RFLP (one allele being present on each chromosome), indicating that neither of her X chromosomes carries this gene deletion in the cells studied. Her carrier risk is not negligible, however, as she may carry a germline mutation that would not be detected by analysing leucocyte DNA. The boy's sister (lane C) is also heterozygous for the associated RFLP (one band inherited from her father and the other from her mother), indicating that she has not inherited a deleted gene from her mother. An inherited mutation would be present in all somatic cells, and heterozygosity for the RFLP in leucocyte DNA confirms that this daughter is not a carrier.

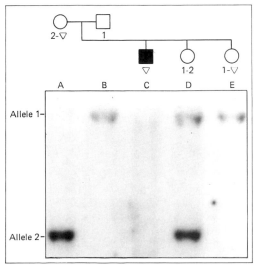

Restriction fragment length polymorphism and deletion detected with probe p20 in DNA digested with *Msp* I restriction endonuclease. (See text for interpretation.) ▽ = Deletion.

Another example of deletion and RFLP analysis is shown in the autoradiograph opposite. The affected boy again has a gene deletion detected by probe p20 (lane C). His mother (lane A) is apparently homozygous for allele 2. One of the boy's sisters (lane E) has inherited an X chromosomal band from her father (lane B) but not her mother. The explanation of this apparent, "non-maternity" is that the daughter has inherited a deleted gene from her mother, and this indicates that they are both carriers. The other sister (lane D) has inherited a paternal and a maternal band. The maternal band corresponds to the non-deleted gene, and this sister is therefore not a carrier.

Molecular genetics of some common mendelian disorders

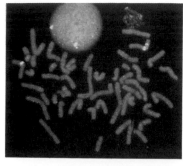

Fluorescence in situ hybridisation in a female carrier of a Duchenne muscular dystrophy mutation involving deletion of exon 47. Hybridisation with a probe from the centromeric region of the X chromosome identifies both X chromosomes. Only one X chromosome shows a fluorescent hybridisation signal with a probe corresponding to exon 47, which indicates that the other X chromosome is deleted for this part of the gene.

An alternative method of identifying gene deletions in females is to use the flurorescence in situ hybridisation (FISH) technique, as described on page 23. A fluorescently labelled DNA probe corresponding to a specific exon will show no hybridisation signal on a chromosome whose dystrophin gene has a deletion of the corresponding exon. For example, a male with Becker's muscular dystrophy caused by a deletion involving exons 45–47 will show no hybridisation with an exon 45 probe. A female carrier of this deletion will show a hybridisation signal on one X chromosome but not on the other. A normal female will show hybridisation signals on both X chromosomes. This provides a direct carrier detection test for female relatives, but as with all other methods of carrier testing, will not exclude the presence of germline mosaicism in the mothers of isolated cases.

In about one third of cases of Duchenne and Becker's muscular dystrophy the mutation is not a large deletion or duplication. Polymorphic DNA markers (either RFLPs or CA repeats) can be used to track the disease gene through these families to make predictions about genetic state in female relatives and male fetuses at risk as shown by the autoradiograph below. The problems associated with this approach include the frequency with which intragenic recombination occurs (12% across the entire gene) and the high incidence of both sporadic cases due to new mutation and of germline mosaicism. The identification of point mutations in cases without deletions, using techniques such as mRNA analysis and single strand conformational polymorphism (SSCP) analysis, allows definitive carrier detection and prenatal diagnosis. The routine application of mutation analysis is currently limited by the complexity of the analysis required, and the likelihood that most point mutations will be unique to an individual family.

DNA analysis has proved to be extremely valuable in investigating families with Duchenne and Becker's muscular dystrophy. Not only can many carriers be identified and offered definitive prenatal diagnosis but female relatives at low risk can be identified and reassured. Several difficulties are still encountered: family studies are required, definitive tests may not be possible if the affected boy is no longer living, and calculations are often complex. In sporadic cases the mother of an affected boy cannot be classed as having a negligible risk because of the possibility of germline mosaicism (calculated to be up to 20%), and even if population screening for carriers becomes possible the disorder will not be eradicated because of the high incidence of new mutations.

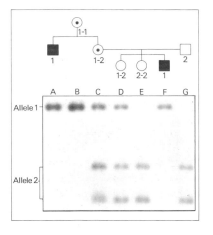

Restriction fragment length polymorphism (alleles 1 and 2) on X chromosome detected by probe pERT 87-15 in DNA digested with *Xmn* I restriction endonuclease. Dystrophy gene segregates with maternal allele 1, indicating that one daughter (lane D) is at high risk of being a carrier and the other (lane E) is at low risk.

Fragile X syndrome

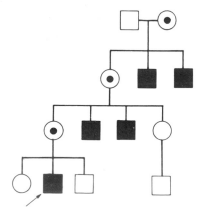

The fragile X syndrome, first described in 1969 and delineated during the mid-1970s, is the most common single cause of severe mental retardation after Down's syndrome. The disorder is estimated to affect around 1 in 2000 individuals, with many more gene carriers. Fragile X syndrome is inherited as an X linked trait, but both males and females can be affected. The clinical phenotype comprises mental retardation of varying degree, macro-orchidism in males, typical facies with prominent forehead, large jaw, and large ears, joint laxity in some cases, and frequent behavioural problems.

X linked recessive pedigree in fragile X syndrome.

Fragile X chromosome.

Mentally retarded brothers with fragile X syndrome.

Chromosomal analysis performed under special culture conditions shows a fragile site near the end of the long arm of the X chromosome at Xq27 in a proportion of cells in most affected males and some affected females, from which this disorder derives its name. The disorder follows X linked inheritance, but unusual features of the condition are the high incidence of retardation in carrier females, the transmission of the disorder by apparently unaffected males (referred to as normal transmitting males), and the increasing severity occurring in successive generations of a family. These observations are now explained by the nature of the fragile X mutation.

The gene for fragile X syndrome has been identified and named FMR-1. The mutation causing fragile X syndrome involves amplification of a CGG trinucleotide repeat within this gene. The number of repeats present in normal genes varies from 6–50. Fragile X mutations can be subdivided into premutations and full mutations. Premutations involve a small increase in the number of CGG repeats ranging between 50 and 200 repeats. Full mutations involve larger expansions, of more than 200 and often several thousand repeats. Premutations do not cause any clinical or cytogenetic abnormality and are found in unaffected carrier females and normal transmitting males.

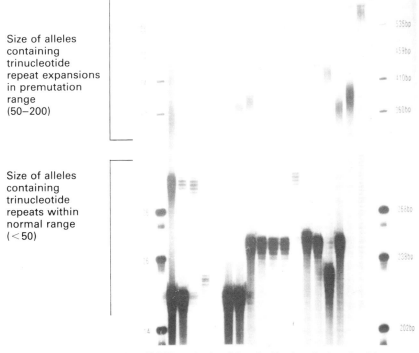

Size of alleles containing trinucleotide repeat expansions in premutation range (50–200)

Size of alleles containing trinucleotide repeats within normal range (<50)

Polymerase chain reaction (PCR) analysis of fragile X related trinucleotide repeat showing examples of alleles with trinucleotide repeats in both normal and premutation size range. (Left axis: number of trinucleotide repeats.) (Right axis: size of amplified allele in base pairs.)

Premutations are unstable when transmitted by a carrier mother, and can become full mutations in both male and female offspring. The full mutation is associated with DNA methylation which inactivates the FMR-1 gene. All males and about 50% of females with the full mutation are mentally retarded. Premutations do not expand when transmitted by a male, and the daughters of normal transmitting males are always unaffected premutation carriers but will be at risk of having affected children themselves.

Before the identification of the underlying mutation, carrier testing and prenatal diagnosis relied mainly on DNA analysis using linked markers and cytogenetic analysis of fetal blood samples. Mutation analysis and estimation of methylation state now enables direct detection of premutation and full mutation carriers within families. In prenatal diagnosis male and female fetuses carrying a premutation are predicted to be clinically unaffected and male fetuses with the full mutation predicted to be clinically affected. The main difficulty arises in cases where a full mutation is identified in a female fetus, as this is associated with a 50% risk of mental retardation which cannot be more clearly defined at present.

The autoradiographs of cystic fibrosis were reproduced by kind permission of Mr A Ivinson and Dr A Wallace; those of Huntington's disease by Dr D Crawford and (of the G8 haplotypes) Professor A Read; those of Duchenne muscular dystrophy by Mr R Mountford and (of the fluorescence in situ hybridisation) Dr Lorraine Gaunt; and that of fragile X syndrome by Dr S Ramsden, all of St Mary's Hospital, Manchester. The illustration showing fragile X Karyotope was reproduced by kind permission of Mr M McKinley, of St Mary's Hospital, Manchester.

GLOSSARY

Alleles — Alternative forms of a gene or DNA sequence occurring at the same locus on homologous chromosomes.

Aneuploid — Chromosome number that is not an exact multiple of the haploid set—for example, $2n-1$ or $2n+1$.

Autoradiography — Detection of radiolabelled molecules with x ray film.

Autosome — Any chromosome other than the sex chromosomes.

Bayesian analysis — Mathematical method for calculating probability of carrier state in mendelian disorders by combining several independent likelihoods.

Carrier — Healthy person possessing a mutant gene in heterozygous form: also refers to a person with a balanced chromosomal translocation.

Chimaera — Presence in a person of two different cell lines derived from fusion of two zygotes.

Chorionic villus sampling — Procedure for obtaining fetally derived chorionic villus material for prenatal diagnosis.

Clone — All cells arising by mitotic division from a single original cell and having the same genetic constitution.

Codominant — Trait resulting from expression of both alleles at a particular locus in heterozygotes—for example, the ABO blood group system.

Codon — Coding sequence of three adjacent nucleotides.

Complementary DNA (cDNA) — Single stranded DNA synthesised from messenger RNA.

Concordance — Presence of the same trait in both members of a pair of twins.

Diploid — Normal state of human somatic cells, containing two haploid sets of chromosomes (2n).

Discordance — Presence of a trait in only one member of a pair of twins.

Dizygotic — Twins produced by the separate fertilisation of two different eggs.

DNA electrophoresis — Separation of DNA restriction fragments by electrophoresis in agarose gel.

DNA fingerprinting — Analysis that detects DNA pattern unique to a given person.

DNA polymerase — Enzyme concerned with synthesis of double stranded DNA from single stranded DNA.

Dominant — Trait expressed in people who are heterozygous for a particular gene.

Dysmorphology — Study of malformations arising from abnormal embryogenesis.

Embryo biopsy — Potential method for preimplantation diagnosis of genetic disorders used in conjunction with in vitro fertilisation.

Empirical risk — Risk of recurrence for multifactorial or polygenic disorders based on family studies.

Exon — Region of a gene transcribed into messenger RNA and translated into protein product.

Fetoscopy — Endoscopic procedure permitting direct visual examination of the fetus.

Genetic counselling — Process by which information on genetic disorders is given to a family.

Genome — Total DNA carried by a gamete.

Genotype — Genetic constitution of an individual person.

Gonadal mosaicism — Presence of a mutation in germline but not somatic cells, which results in transmission of a genetic disorder by a healthy person.

Haploid — Normal state of gametes, containing one set of chromosomes (n).

Haplotype — Particular set of alleles at closely linked loci on a single chromosome that are inherited together.

Hemizygote — Describes the genotype of males with an X linked trait, as males have only one X chromosome.

Heritability — The contribution of genetic as opposed to environmental factors to phenotypic variance.

Heterozygote — Person possessing different alleles at a particular locus on homologous chromosomes.

Holandric inheritance — Pattern of inheritance of genes on the Y inheritance chromosome.

Homologous chromosomes — Chromosomes that pair at meiosis and contain the same set of gene loci.

Homozygote — Person having two identical alleles at a particular locus on homologous chromosomes.

Hybridisation — Process by which single strands of DNA with homologous sequences bind together.

Intron — Region of a gene transcribed into messenger RNA but spliced out before translation into protein product.

Karyotype — Description of the chromosomes present in somatic cells.

Kilobase (kb) — 1000 Base pairs (bp) of DNA.

Linkage — Term describing genes or DNA sequences situated close together on the same chromosome that tend to segregate together.

Linkage disequilibrium — Occurrence together of two particular linked alleles on the same chromosome more commonly than expected by chance.

Locus — Site of a specific gene or DNA sequence on a chromosome.

Lyonisation — Process of X chromosome inactivation in cells with more than one X chromosome.

Marker — General term for biochemical or DNA polymorphism occurring close to a gene and used in gene tracking.

Meiosis — Cell division during gametogenesis resulting in haploid gametes.

Mendelian disorder — Inherited disorder due to a defect in a single gene.

Mitosis — Cell division occurring in somatic cells resulting in diploid daughter cells.

Monosomy — Loss of one of a pair of homologous chromosomes.

Monozygotic — Twins derived from a single fertilised egg.

Mosaic — Presence in a person of two different cell lines derived from a single zygote.

Multifactorial inheritance — Disorder caused by interaction of more than one gene plus the effect of environment.

Multiple alleles — Existence of more than two alleles at a particular locus.

Mutation — Change in the structure of DNA.

Non-dysjunction — Failure of separation of paired chromosomes during cell division.

Obligate carrier — Family member who must be a heterozygous gene carrier, determined from the mode of inheritance and the pattern of affected relatives within the family.

Oligoprobe — A short DNA probe whose hybridisation is sensitive to a single base mismatch.

Oncogene — Gene with potential to cause cancer.

Penetrance — Probability that a disease genotype will result in an abnormal phenotype.

Phenotype — Physical or biochemical characteristics of a person reflecting genetic constitution and environmental influence.

Point mutation — Substitution of a single base pair in DNA molecule that may affect protein synthesis.

Polygenic inheritance — Disorder caused by interaction between more than one gene.

Polymerase chain reaction (PCR) — Method of amplification of specific DNA sequences by repeated cycles of DNA synthesis to permit rapid analysis of DNA restriction fragments subsequently.

Polymorphism	Genetic characteristic with more than one common form in a population.
Polyploid	Chromosome numbers representing multiples of the haploid set greater than diploid—for example, 3n.
Polysome	Group of ribosomes associated with a particular messenger RNA molecule.
Post-translational modification	Alterations to protein structure after synthesis.
Proband	Index case through which a family is identified.
Probe	Labelled DNA fragment used to detect complementary sequences in DNA sample.
Pseudogene	Functionless copy of a known gene.
Pulse field gel electrophoresis	Method for separating large fragments of DNA (50–10 000 kb) by altering the direction of the electrical field during electrophoresis.
Purine	Nitrogenous base: adenine or guanine.
Pyrimidine	Nitrogenous base: cytosine, thymine, or uracil.
Recessive	Trait expressed in people who are homozygous or hemizygous for a particular gene but not in those who are heterozygous for the gene.
Recombination	Crossing over between homologous chromosomes at meiosis which separates linked loci.
Restriction endonuclease	Enzyme that cleaves double stranded DNA at a specific sequence.
Restriction fragments	DNA fragments produced by restriction endonuclease digestion of sample DNA.
Restriction fragment length polymorphism (RFLP)	Variation in size of DNA fragments produced by restriction endonuclease digestion due to variation in DNA sequence at the enzyme recognition site.
Reverse transcriptase	Enzyme catalysing the synthesis of complementary DNA from messenger RNA.
Segregation	Separation of alleles during meiosis so that each gamete contains only one member of each pair of alleles.
Southern blotting	Process of transferring DNA fragments from agarose gel on to nitrocellulose filter or nylon membrane.
Splicing	Removal of introns and joining of exons in messenger RNA.
Trait	Recognisable phenotype due to a genetic character.
Transcription	Production of messenger RNA from DNA sequence in gene.
Translation	Production of protein from messenger RNA sequence.
Translocation	Transfer of chromosomal material between two non-homologous chromosomes.
Triploid	Cells containing three haploid sets of chromosomes (3n).
Trisomy	Cells containing one more than the normal diploid set of chromosomes (2n+1).
Unifactorial	Inheritance controlled by single gene pair.

SUPPORT GROUPS

There are support groups for families with many different genetic disorders. The following is intended as a guide and does not include all existing self help groups. The *Contact a Family Directory of Specific Conditions and Rare Syndromes in Children* contains details of support networks on over 200 disorders and is available from Contact a Family, 170 Tottenham Court Road, London W1P CHA.

In the UK the Genetic Interest Group (GIG) c/o Institute of Molecular Medicine, John Radcliffe Hospital, Oxford co-ordinates the activities of the different societies and lobbies the government and other bodies for improved services.

Association for Spina Bifida and
  Hydrocephalus
42 Park Road
Peterborough
Cambridgeshire PE1 2UQ

The Arthrogryposis Group (TAG)
1 The Oaks
Gillingham
Dorset SP8 4SW

The British Retinitis Pigmentosa Society
Pond House
Lillingstone
Dayrell
Bucks MK18 5AS

Brittle Bone Society
122 City Road
Dundee DD2 2PW

Child Growth Foundation
2 Mayfield Avenue
Chiswick
London W4 1PW

Cleft Lip and Palate Association
1 Eastwood Gardens
Kenton
Newcastle upon Tyne NE3 3DQ

The Cystic Fibrosis Trust
Alexandra House
5 Blyth Road
Bromley
Kent BR1 3RS

Down's Syndrome Association
155 Mitcham Road
London SW17 9PG

Dystrophic Epidermolysis Bullosa Research
  Association (DEBRA)
Debra House
13 Wellington Business Park
Duke's Ride
Crowthorne
Berkshire RG45 6LS

Fragile X Society
53 Winchelsea Lane
Hastings
East Sussex TN35 4LG

The Friedreich's Ataxia Group
Copse Edge
Thursley Road
Elstead
Godalming
Surrey GU8 6DJ

The Haemophilia Society
123 Westminster Bridge Road
London SE1 7HR

Huntington's Disease Association
108 Battersea High Street
London SW11 3HP

Marfan Association UK
6 Queens Road
Farnborough
Hants GU14 6DH

Muscular Dystrophy Group of Great Britain
7–11 Prescott Place
London SW4 6BS

National Deaf Children's Society
24 Wakefield Road
Leeds LS26 0SF

National Society for Phenylketonuria UK
7 Southfield Close
Willen
Milton Keynes
Bucks MK15 9LL

The Neurofibromatosis Association
82 London Road
Kingston on Thames
Surrey KT2 6QJ

Research Trust for Metabolic Diseases in
  Children (RTMDC)
Golden Gates Lodge
Weston Road
Crewe
Cheshire CW1 1XN

Restricted Growth Associaton
170 Tottenham Court Road
London W1P CHA

Royal National Institute for the Blind
224 Great Portland Street
London W1N 6AA

Royal National Institute for the Deaf
105 Gower Street
London WC1E 6AH

Sickle Cell Society
54 Station Road
London NW10 4UA

The Society for Mucopolysaccharide
  Diseases
55 Hill Avenue
Amersham
HP6 5BX

Stillbirths and Neonatal Deaths Society
  (SANDS)
28 Portland Place
London W1N 4DE

Tuberous Sclerosis Association of Great
  Britain
Little Barnsley Farm
Milton Road
Catshill
Bromsgrove
Worcestershire B61 0WQ

UK Rett Syndrome Association
29 Carlton Road
Friern Barnet
London N11 3EX

UK Thalassaemia Society
107 Nightingale Lane
London N8 7QY

FURTHER READING LIST

This list does not aim to be exhaustive, but to give examples of books recommended by the author for further reading and reference. Additional reading suggestions giving wider coverage of different organ systems are included in Peter S Harper's excellent book *Practical Genetic Counselling*.

Undergraduate/introductory books

Connor J M, Ferguson-Smith M A. *Essential medical genetics*. 4th ed. Oxford: Blackwell, 1993.

Emery A E, Mueller R F. *Elements of medical genetics*. 8th ed. Edinburgh: Churchill Livingstone, 1992.

Gelehrter T D, Collins F S. *Principles of medical genetics*. Baltimore: Williams and Wilkins, 1990.

Read A. *Medical genetics: an illustrated outline*. London: Gower Medical, 1989.

Short texts

Harper P S. *Practical genetic counselling*. 4th ed. London: Wright, 1993.

Weatherall, D J. *The New Genetics and Clinical Practice*. 3rd ed. Oxford: Oxford University Press, 1991.

Fuhrmann W, Vogel F. *Genetic counselling: a guide for the practising physician*. 2nd ed. Berlin: Springer, 1983.

Strachan T. *The human genome*. Oxford: Bios Scientific Publishers, 1992.

Young I D. *Introduction to risk calculation in genetic counselling*. Oxford: Oxford University Press, 1991.

Jones K L. *Smith's recognisable patterns of human malformation*. 4th ed. Philadelphia: W B Saunders, 1988.

Baraitser M, Winter R. *A colour atlas of clinical genetics*. London: Wolfe, 1988.

Baraitser M. *The genetics of neurological disorders*. Oxford: Oxford University Press.

Bundey S. *Genetics and neurology*. 2nd ed. Edinburgh: Churchill Livingstone, 1992.

Clarke A. ed. *Genetic counselling. Practice and principles*. London: Routledge, 1994.

Databases

Winter R M, Baraitser M. *London dysmorphology database and dysmorphology photo library*. Oxford: Oxford University Press, 1993.

Baraitser M, Winter R M. *London neurogenetics database*. Oxford: Oxford University Press, 1993.

Bankier A. *POSSUM (dysmorphology, database and photo library)*. Melbourne, Australia: Murdoch Institute.

Bankier A. *OSSUM (skeletal dysplasia database and photo library)*. Melbourne, Australia: Murdoch Institute.

Reference texts

Rimoin D L, Connor J M, Pyeritz R E, Emery A E H, eds. *Emery and Rimoin's principles and practice of medical genetics*. Edinburgh: Churchill Livingstone, 1994.

McKusick V A. *Mendelian inheritance in man*. 10th ed. Baltimore: Johns Hopkins, 1992. (Also available on line).

Vogel F, Motulsky A G. *Human genetics, problems and approaches*. 2nd ed. Berlin: Springer, 1986.

Gorlin R J, Cohen M M, Levin L S. *Syndromes of the head and neck*. 3rd ed. Oxford: Oxford University Press, 1990.

Schinzel A. *Catalogue of unbalanced chromosome aberrations in man*. Berlin: De Gruyter, 1983.

De Grouchy, J, Turleau, C. *Clinical atlas of human chromosomes*. New York: Wiley, 1982.

Scriver. *Metabolic basis of inherited disease*. New York: McGraw-Hill.

Brock D J H, Rodeck C H, Ferguson-Smith M A. *Prenatal diagnosis and screening*. Edinburgh: Churchill Livingstone, 1992.

Shepard T H. *Catalog of teratogenic agents*. 7th ed. Baltimore: Johns Hopkins, 1992.

Grudzinskas J G, Chard T, Chapman M, Cuckle H, eds. *Screening for Down's syndrome*. Cambridge: Cambridge University Press, 1994.

Strachan T, Read A P. *Human molecular genetics*. Oxford: Bios Scientific Publishers, 1996.

Evans M I, ed. *Reproductive risks and prenatal diagnosis*. Connecticut: Appleton and Lange, 1992.

Stevenson R E, Hall J G, Goodman R M, eds. *Human malformations and related anomalies*. New York: Oxford University Press, 1993.

Donnai D, Winter R M, eds. *Congenital malformation syndromes*. London: Chapman and Hall Medical, 1995.

Winter R M, Knowles S A S, Bieber F R, Baraitser M. *The malformed fetus and stillbirth. A diagnostic approach*. Chichester: John Wiley and Sons, 1988.

INDEX

Index

ABCs *from the* BMJ

ABC OF ALCOHOL
THIRD EDITION
Edited by Alex Paton
BMJ

0 7279 0812 X

ABC OF ASTHMA
THIRD EDITION
John Rees
John Price
BMJ

0 7279 0882 0

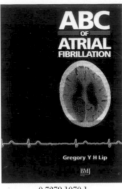

ABC OF ATRIAL FIBRILLATION
Gregory Y H Lip
BMJ

0 7279 1070 1

ABC OF BRAINSTEM DEATH
C PALLIS and DH HARLEY
SECOND EDITION

0 7279 0245 8

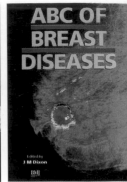

ABC OF BREAST DISEASES
Edited by J M Dixon
BMJ

0 7279 0915 0

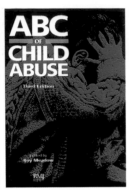

ABC OF CHILD ABUSE
Third Edition
Edited by Roy Meadow
BMJ

0 7279 1106 6

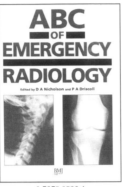

ABC OF EMERGENCY RADIOLOGY
Edited by D A Nicholson and P A Driscoll
BMJ

0 7279 0832 4

ABC OF EYES
SECOND EDITION
P T KHAW
A R ELKINGTON
BMJ

0 7279 0766 2

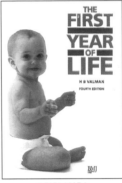

THE FIRST YEAR OF LIFE
H B VALMAN
FOURTH EDITION

0 7279 0897 9

ABC OF HYPERTENSION
THIRD EDITION
E O'Brien, D G Beevers, and H M Marshall
BMJ

0 7279 0769 7

ABC OF HEALTHY TRAVEL
Fifth Edition
BMJ

0 7279 1138 4

ABC OF MAJOR TRAUMA
Second Edition
Edited by David Skinner, Peter Driscoll, Richard Earlam
BMJ

0 7279 0917 7

ABC OF MEDICAL COMPUTING
Nicholas Lee and Andrew Millman
BMJ

0 7279 1046 9

ABC OF RESUSCITATION
M COLQUHOUN, A HANDLEY, T EVANS
THIRD EDITION

0 7279 0839 1

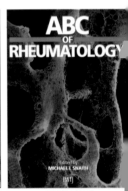

ABC OF RHEUMATOLOGY
Edited by MICHAEL L SNAITH
BMJ

0 7279 0997 5

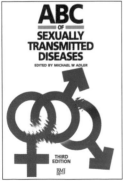

ABC OF SEXUALLY TRANSMITTED DISEASES
EDITED BY MICHAEL W ADLER
THIRD EDITION
BMJ

0 7279 0889 8

ABC OF SLEEP DISORDERS
Edited by Colin M Shapiro
BMJ

0 7279 0794 8

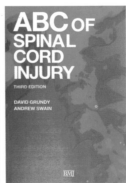

ABC OF SPINAL CORD INJURY
THIRD EDITION
DAVID GRUNDY
ANDREW SWAIN
BMJ

0 7279 1049 3

ABC OF SPORTS MEDICINE

0 7279 0844 8

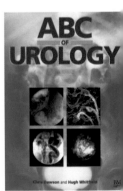

ABC OF UROLOGY
Chris Dawson and Hugh Whitfield
BMJ

0 7279 1075 2

For further details contact your local bookseller or write to:

BMJ Publishing Group
BMA House
Tavistock Square
London WC1H 9JR (U.K.)

BMJ
Publishing
Group